Infrared Thermography

Errors and Uncertainties

Infrared Thermography

Errors and Uncertainties

Waldemar Minkina and Sebastian Dudzik

Częstochowa University of Technology, Poland

WILEY

A John Wiley and Sons, Ltd, Publication

This edition first published 2009

Registered office
John Wiley & Sons Ltd, The Atrium, Southern Gate, Chichester, West Sussex, PO19 8SQ, United Kingdom

For details of our global editorial offices, for customer services and for information about how to apply for permission to reuse the copyright material in this book please see our website at www.wiley.com.

Library of Congress Cataloguing-in-Publication Data

Minkina, Waldemar.
Infrared thermography : errors and uncertainties / Waldemar Minkina and Sebastian Dudzik.
p. cm.
Includes bibliographical references and index.
ISBN 978-0-470-74718-6 (cloth)
1. Thermography. 2. Infrared imaging. 3. Uncertainty. 4. Tolerance (Engineering) I. Dudzik, Sebastian, 1975- II. Title.
TA1570.M62 2009
621.36'2–dc22

2009031444

A catalogue record for this book is available from the British Library.

ISBN: 978-0-470-74718-6 (Hbk)

Set in 10/12pt, Times by Thomson Digital, Noida, India.
Printed in Great Britain by CPI Antony Rowe, Chippenham, Wiltshire.

Man, being the servant and interpreter of Nature, can do and understand so much and so much only as he has observed in fact or in thought of the course of nature. Beyond this he neither knows anything nor can do anything.

F. Bacon, *Novum Organum*, Aphor. I

To our wives

Elżbieta and Barbara

Contents

Preface

In our contact with users of infrared systems we were frequently asked, 'How do you estimate the accuracy of infrared thermography measurements, or how accurate are the data used from thermography measurements, for example, in the analysis of the temperature field of selected objects by the finite difference method (FDM), the finite element method (FEM) or the boundary element method (BEM) (Özisik, 1994, Minkina, 1994, Minkina, 1995, Minkina, 2004, Astarita *et al.*, 2000, Hutton, 2003)?'

The answer to such a question is not straightforward, so we decided to write this book, which is intended to deal with the problem in depth. It is worth underlining that the problem has not yet been fully solved in the literature. Authors, be they physicists, architects, mechanical engineers, power engineers or computer scientists, describe it in different ways, depending on the scientific field they represent. In this monograph we deal with the problem comprehensively, in accordance with international recommendations as published in the *Guide to the Expression of Uncertainty in Measurement* (Guide, 1995, Guide, 2004). This work is the first to deal with the issue in this manner. It is an extension and complement of the study presented in §10 of the monograph by Minkina (2004).

This book also aims to explain the many misunderstandings in the interpretation of temperature measurements and feasible metrological evaluation of commercially available infrared systems.

The first misunderstanding is the wrong interpretation of the Noise Equivalent Temperature Difference (NETD) parameter, published in catalogs as thermal sensitivity and interpreted sometimes as a parameter related to the precision of an infrared thermography measurement. In fact, the NETD parameter is rather for marketing purposes and says little about the actual error of a measurement. This parameter has an effect only on the quality of a thermogram, because it guarantees better uniformity of signals acquired from the particular detectors of the detector array. In practice, it can only give information on the error of temperature difference between two points of a given area of uniform emissivity, measured by the same pixel of a multipixel array (matrix) of detectors in idealized measurement conditions of short camera-to-object distance and no external sources emitting disturbing radiation. It takes place when the measurement model stored in the camera's microcontroller memory is fulfilled and the model parameters (ε_{ob}, T_{atm}, T_o, ω, d) are entered with zero error. Of course, it is difficult to conduct such a measurement in reality.

The second misunderstanding is the wrong interpretation of another parameter published in catalogs: namely, the accuracy of a thermography measurement. This accuracy is associated firstly with the quality of calibration of the array detector (Minkina, 2004). The better the calibration (i.e. the more accurate the bringing of the static characteristics of individual detectors to the same common shape), the smaller the measurement error. Secondly, the measurement accuracy is affected by calibration conducted by the camera manufacturer. Parameters (*R*, *B*, *F*) of the static characteristic of the measurement path determined during calibration are obviously burdened with errors. Therefore, if in a catalog this error is given as ±2 °C, ±2%, then for a given measurement range the larger of the two values should be taken. For example, for a measurement range of 0–100 °C we should take ±2 °C, while for a range of 100–500 °C we should take ±2%. As before, the error value refers to idealized measurement conditions: that is, an adequate measurement model stored in the microcontroller memory and zero errors in the entered model parameters. Under actual conditions (e.g. for a long camera-to-object distance or in the presence of external radiation interfering with the object radiation), the error can be many times greater. In extremely difficult atmospheric conditions, non-contact temperature measurement is not possible at all.

The uncertainty analysis of thermography measurement using analytic methods is very difficult because it involves a complex form of the model (Dudzik, 2005, Minkina, 2004). Therefore, for the uncertainty analysis of the processing algorithm in this work, we use the numerical method for the propagation of distributions recommended by Working Group No. 1 of the BIPM (International Bureau of Weights and Measures) (Guide, 2004). The uncertainty analysis was carried out for correlated as well as for uncorrelated model input variables. It allowed for quantitative evaluation of the influence of individual factors on the expanded uncertainty of infrared camera temperature measurement.

From a terminology perspective, this can be explained using various concepts. In the literature, besides 'thermovision' the term 'thermography' is often used. As the measurements are often computerized, the term 'computer-aided thermography' is used as well. 'Thermography' can be understood as the older technique (e.g. the recording of thermal images on heat-sensitive paper with a thermograph). In this method, firstly the image is obtained and next, observations are taken. Additionally, 'thermography' suggests that we describe graphic systems rather than vision systems. In the English literature, 'computer-aided thermography' is often used. Contemporary thermal imaging systems are called infrared cameras. Sometimes they can be called thermographs as well. Therefore, it seems that the terms 'thermography' and 'thermovision' can be treated interchangeably; in this book, however, the first of these terms is preferred.

The material presented is divided into six chapters. Chapter 1 gives the reader an introduction to the theory of error and uncertainty. Chapter 2 deals with the basic issues of measurements in infrared thermography, such as the law of heat exchange by radiation and emissivity. In Chapter 3 we describe a typical processing algorithm of the measurement path as well as a generalized model of the temperature measurement of the example of FLIR's ThermaCAM PM 595 LW infrared camera.

It is necessary to emphasize that, for other types of infrared cameras and manufacturers, the results and conclusions will be very similar.

Chapter 4 deals with the issue of the measurement error analysis of an infrared system, performed using classic methods. In Chapter 5 we describe the results of simulation research on the uncertainty in measurement in the infrared thermography obtained, using numerical methods for the propagation of distributions.

Waldemar Minkina and Sebastian Dudzik
Częstochowa, 2009

About the Authors

Waldemar Andrzej Minkina was born in 1953 in Częstochowa, Poland. In 1977 he graduated from the Faculty of Electrical Engineering of Częstochowa University of Technology, specializing in the automatization of electric drives. He received a first class honors Ph.D. degree in 1983 from the Institute of Electrical Metrology at Wrocław University of Technology, Poland, and a D.Sc. (habilitation) degree in 1995 from the Faculty of Automatic Control at Lwów Technical University, Ukraine, recommended by the Chair of Measurement and Information Techniques. On 22 June 2006, the President of Poland presented him with a professorial nomination in technical sciences (full professor).

Professor Minkina's research interests include thermometry, computerized thermography, heat measurements and theory, and the techniques of heat exchange. He is the author or co-author of four monographs in metrology: *Measurements of thermal parameters of heat-insulating materials – methods and instruments* (in Polish), Częstochowa University of Technology Publishers, 2004 (ISBN 83-7193-216-2); *Thermovision measurements – methods and instruments* (in Polish), Częstochowa University of Technology Publishers, 2004 (ISBN 83-7193-237-5); *Compensation of dynamic characteristics of thermometric sensors – methods, systems, algorithms* (in Polish), Częstochowa University of Technology Publishers, 2004 (ISBN 83-7193-243-X); and *Thermovision measurements in practice* (in Polish), PAK Agenda Publishers, Warsaw 2004 (ISBN 83-87982-26-1). He has also published 110 journal papers (including 25 published, mainly as the single author, in *Sensors and Actuators*, *Measurement*, *Technisches Messen*, *Experimental Technique of Physics*, *IVUZ Priborostroenije*, *Messen-Pruefen-Automatisieren*, *Messen-Steuern-Regeln*, *Metrology and Measurement Systems* and *The Archive of Mechanical Engineering*). He is an author of six patents, four patent announcements and supervisor of three Ph.D. theses defended with honors.

Professor Minkina has been a visiting professor to institutes of metrology at the Universities of Karlsruhe, West Berlin, Sankt Petersburg and Lviv, as well as in the Physikalisch-Technische Bundesanstalt (PTB) in Berlin and in Risø National Laboratory, Denmark. He was a guest lecturer for Ph.D. studies conducted in the Institute of Solid-State Electronics at Dresden Technical University. He closely collaborates with the Chair of Metrology at Rostock University in the field of computerized thermography. The results of this collaboration are the International Workshops 'Infrarot – Thermografie'.

Professor Minkina is a Member of the Instrumentation and Measurement Systems Section of the Committee of Measurement and Scientific Instrumentation of the Polish Academy of Sciences; a Member of the Program Committee of the monthly journal *Pomiary Automatyka Kontrola* (*Measurement, Automation and Monitoring*) and editor of the Thermometry section; a Member of the Polish Association of Sensor Technology, the Polish Association of Theoretical and Applied Electrotechnics and the Association of Polish Electricians, where he is an expert in three fields. He has served as a Member of the Program, Scientific and Organization Committees of many international and national conferences and many times as a reviewer of journal papers submitted for publication. He was also a reviewer of many grants and projects conducted by the State Committee for Scientific Research (KBN). Since 1996 he has held the Chair of Microprocessor Systems, Automatic Control and Heat Measurements. In 1995–2005 he became the Director of the Institute of Electronics and Control Systems.

Sebastian Dudzik was born in 1975 in Łódź, Poland. In 2000 he graduated from the Faculty of Electrical Engineering at Częstochowa University of Technology, specializing in measurement and control systems. Since 2000 he has been in the Faculty of Electrical Engineering at Częstochowa University of Technology, where he received his Ph.D. degree in technical sciences in 2007. He is the author or co-author of 21 papers published in journals and conference proceedings in both Poland and abroad. His research interests include the applications of active infrared thermography, artificial neural networks and neuro-fuzzy models of heat exchange and non-destructive testing.

Acknowledgements

We would like to cordially thank the five reviewers of this book. The remarks in their reviews have considerably improved the contents of this publication.

We would also like to thank Dr Janusz Baran for translating the text from Polish to English.

Symbols

A	absorbance coefficient
α	angle of observation, rad
B	one of three calibration constants of the infrared camera (the others being F, R)
$C_o = (5.670\,32 \pm 0.000\,71)\cdot 10^8\ \mathrm{W\,m^{-2}\,K^{-4}}$	technical constant of black body radiation (ISO 31)
$c = 299\,792\,458 \pm 1.2\ \mathrm{m\,s^{-1}}$	speed of light in vacuum (ISO 31)
$c_1 = 2\cdot\pi\cdot h\cdot c^2 = (3.741\,832 \pm 0.000\,020)\cdot 10^{-16}\ \mathrm{W\,m^2}$	first radiant constant (ISO 31)
$c_2 = h\cdot c/k = (1.438\,786 \pm 0.000\,045)\cdot 10^{-2}\ \mathrm{m\,K}$	second radiant constant (ISO 31)
$D^*(\lambda, T)$	normalized spectral detectivity, $\mathrm{cm\,Hz^{1/2}\,W^{-1}}$
d	camera-to-object distance (one of the input variables in the infrared camera model), m
$\Delta_{T_{ob}}$	absolute error of a measurement model in infrared thermography, K or °C
Δf	frequency bandwidth, Hz
$\delta_{T_{ob}}$	relative error of a measurement model in infrared thermography
$E(X)$	expected value of a discrete random variable X
ε	emissivity (one of the input variables in the infrared camera model)
F	area, $\mathrm{m^2}$ (one of three calibration constants of the infrared camera, the others being B, R)
Φ	heat flux, W; power density of thermal radiation, $\mathrm{W\,m^{-2}}$
$h = (6.626\,176 \pm 0.000\,036)\cdot 10^{-34}\ \mathrm{W\,s^2}$	the Planck constant (ISO 31)
I_v	luminous intensity, cd
$k = (1.380\,662 \pm 0.000\,044)\cdot 10^{-23}\ \mathrm{W\,s\,K^{-1}}$	the Boltzmann constant (ISO 31); expansion factor
L_v	luminance, $\mathrm{cd\,m^{-2}}$
λ	wavelength, μm
M	radiant exitance, $\mathrm{W\,m^{-2}}$
q	thermal flux density, $\mathrm{W\,m^{-2}}$

R — reflectance (reflectivity) coefficient (one of three calibration constants of the infrared camera, the others being B, F); outer radius, m

ρ — correlation coefficient among input variables of infrared camera measurement model

$S_k(\lambda)$ — function describing relative spectral sensitivity of the camera

s_{ob} — output signal from a detector, corresponding to the object temperature

$\sigma(X)$ — standard deviation of a random variable X

$\sigma_o = 2 \cdot \pi^5 \cdot k^4/(15 \cdot h^3 \cdot c^2) = (5.670\,32 \pm 0.000\,71) \cdot 10^{-8}\ \mathrm{W\,m^{-2}\,K^{-4}}$ — the Stefan–Boltzmann constant (ISO 31)

TT — transmission coefficient

T_{ob} — temperature of object, K or °C

T_o — ambient temperature, K or °C

T_{atm} — atmospheric temperature, K or °C

$u(x_i)$ — standard uncertainty of ith input variable in the infrared camera model

$u_c(T_{ob})$ — combined standard uncertainty of object temperature

ω — humidity (one of the input variables in the infrared camera model), %

Glossary

Absolute error of a measurement is the difference between measured value $\hat{y}$ and actual value y.

Absolute error of a measurement model in infrared thermography is the difference between value T_C calculated by the camera measurement path algorithm for a single element (pixel) of the array detector and actual temperature T_R of the surface area mapped (represented) by this element.

Accuracy (of measurement) is a maximum deviation, expressed as % of scale or in degrees Celsius, that the reading of an instrument will deviate from a correct standard reference.

Black body, black body radiator is a body that absorbs all incident radiation. From Kirchhoff's law it follows that a black body is also a perfect radiator. The emissivity of a black body is equal to one.

Bolometric detectors are resistors of very small heat capacity with a large, negative temperature coefficient of resistivity.

Calibration is a procedure for checking and/or adjusting an instrument. After calibration, the readings of the instrument will agree with a standard. Calibration removes instrument systematic error but is not able to remove random errors.

Combined standard uncertainty $u_c(y)$ is the positive square root of the combined variance $u_c^2(y)$, defined as $u_c^2(y) = \sum_{i=1}^{N} (\partial f/\partial x_i)^2 u^2(x_i)$, where $y = f(x_1, x_2, \ldots, x_n)$ is the measurement model function and $u^2(x_i)$ is the variance of the ith input of the model.

Confidence level$(1-\alpha)$ is a value of probability associated with a confidence interval or statistical coverage interval.

Data processing algorithm uncertainty is a measure of the spread of an output random variable, equal to the standard experimental deviation of this variable.

Emissivity ε of a body for the full radiation range, called the total emissivity, is the ratio of full-range radiant exitance of that body to full-range radiant exitance of a black body at the same temperature.

Expanded uncertainty U is the uncertainty obtained by multiplying combined standard uncertainty $u_c(y)$ by expansion factor k: $U = ku_c(y)$.

Expected value $E(X)$ of a discrete random variable X, whose values x_i appear with probabilities p_i, is $E(X) = \sum p_i x_i$.

Field of view (FOV) is an area that can be observed from a given distance d using the optics installed on an infrared camera.

Gray body is an object whose emissivity is a constant value less than unity over a specific spectral range.

Instantaneous field of view (IFOV) is the field of view of a single detector (pixel) in a detector array.
Limiting error is the smallest range around the measured value $\hat{y}$ containing actual value y.
Luminance or ***brightness*** $\boldsymbol{L_v}$ is the surface density of luminous intensity in a given direction.
Luminous intensity $\boldsymbol{I_v}$ is the light flux in a given direction per unit solid angle.
Method of increments (exact method) consists of determining the increment of a measurement model function for the known increments of input quantities (i.e. absolute errors).
Method of total differential (approximated method) is based on the expansion of a measurement model function in a Taylor series around the point defined by the actual (true conventional) values of the inputs.
Monochromatic emissivity $\boldsymbol{\varepsilon_\lambda}$ is the ratio of monochromatic radiant exitance $M_\lambda(\lambda,T)$ of a body at a given wavelength λ to monochromatic radiant exitance $M_{b\lambda}(\lambda,T)$ of a black body at the same wavelength, the same temperature and observed at the same angle.
Noise equivalent power(NEP) is the RMS (Root Mean Square) power of incident monochromatic radiation of wavelength λ that generates an output voltage whose RMS value is equal to the level of noise normalized to unit bandwidth.
Noise equivalent temperature difference (NETD) is the difference between the temperature of an observed object and the ambient temperature that generates a signal level equal to the noise level.
Non-gray body is an object whose emissivity varies with wavelength over the wavelength interval of interest.
One-sided coverage interval: if T is a function of observed values, such that for estimated parameter of population θ, probability $Pr(T \geq \theta)$ or $Pr(T \leq \theta)$ is at least equal to $(1-\alpha)$ (where $(1-\alpha)$ is a fixed number, positive and smaller than one), then the interval from the smallest possible value of θ to T (or the interval from T to the biggest possible value of θ) is the one-sided coverage interval θ with confidence level $(1-\alpha)$.
Pyroelectric detectors are built from semiconductors that exhibit the so-called pyroelectric effect.
Quantile of order $\boldsymbol{\beta}$ of a probability distribution described by cumulative distribution function $G(\eta)$ is such that, for a value η of the random variable, equality $G(\eta) = \beta$ is satisfied. This means that the probability of occurrence of this value is equal to β.
Radiant exitance (emittance) is the ratio of (temperature- and wavelength-dependent) radiant power (radiant flux) $d\Phi$ emitted by an arbitrarily small element of surface containing the considered point to a projected area dF of that element.
Radiant intensity is the radiant flux per unit solid angle.
Random error is the difference between the result of an individual measurement and the mean value calculated for an infinite number of measurements of a quantity, carried out under the same conditions.
Relative error of a measurement is the ratio of the absolute error to the actual value.
Relative error of a measurement model in infrared thermography is the ratio of absolute error $\Delta_{T_{ob}}$ to actual temperature T_R.
Response rate is a parameter determined by the detector's time constant.
Slit response function (SRF) is a parameter that, similar to IFOV, describes the capability of a camera with an array detector to measure the temperature of small objects.
Standard deviation $\sigma(X)$ of a random variable is the positive square root of the variance.

Standard uncertainty of a measurement is the uncertainty of that measurement expressed in the form of the standard deviation.

Systematic error (bias) is the difference between the mean value calculated for an infinite number of measurements of a quantity – carried out under the same conditions – and its actual value.

Temperature sensitivity is a parameter that determines change of signal per unit change of temperature for object temperature $T_{ob} = T_o$.

Thermopile detectors are built as a thermopile, that is a system of thermoelements connected in series.

Type A standard uncertainty is the standard uncertainty determined on the basis of the observed frequency distribution.

Type B standard uncertainty is the standard uncertainty determined on the basis of a frequency distribution assumed a priori.

Uncertainty of a measurement is a parameter characterizing the spread of measurement values that can be assigned to the measured quantity in a justified way.

Voltage or current (spectral) sensitivity is a ratio of the RMS value of the first harmonic of a detector output voltage (current) to the RMS value of the first harmonic of incident radiation power.

1

Basic Concepts in the Theory of Errors and Uncertainties

1.1 Systematic and Random Errors

In modern measurement systems we can observe, along with growth of complexity, the evolution of measurement methods to estimate accuracy. On one hand, this is a consequence of the increasing complexity of measurement models: the number of input quantities increases and dependencies between inputs and outputs become more and more complicated. It makes it difficult to estimate accuracy with the use of classical methods that employ analytical descriptions. On the other hand, technical progress enables better insight into physical reality, which, among other things, involves changes to definitions of units of measure, which are the basis of each metric system. For example, consider how the definition of the meter has evolved over the last two centuries (www.gum.gov.pl):

1793: The meter is 1/10 000 000 of the distance from the equator to the Earth's North Pole (i.e. the Earth's circumference is equal to 40 million meters).
1899: The meter is the distance, measured at 0 °C, between two engraved lines on the top surface of the international prototype meter standard, made of a platinum–iridium bar (102 cm in length) with an H-shaped cross-section.
1960: The meter is equal to 1 650 763.73 wavelengths of the orange–red radiation of the krypton-86 isotope.
1983: The meter is the distance traveled by light in vacuum in 1/299 792 458 seconds.

For the evaluation of measurement accuracy, it is necessary to define basic theoretical concepts of error and uncertainty. Below we present definitions of the measurement error for a single value of a measured quantity.

The **absolute error** of a measurement is the difference between measured value $\hat{y}$ and actual value y:

$$\Delta y = \hat{y} - y. \tag{1.1}$$

Infrared Thermography: Errors and Uncertainties Waldemar Minkina and Sebastian Dudzik

The **relative error** of a measurement is the ratio of the absolute error to the actual value:

$$\delta y = \frac{\Delta y}{y} = \frac{\hat{y} - y}{y}. \quad (1.2)$$

In practice, the unknown actual value y in formula (1.1) is substituted by the true conventional value. Since the exact value of the absolute error is unknown, it is very important to evaluate a range in which the actual value is located. Such reasoning leads to the definition of the limiting error.

The **limiting error** is the smallest range around the measured value $\hat{y}$ containing actual value y (Guide 2004):

$$\hat{y} - \Delta y_{\min} \geq y \geq \hat{y} + \Delta y_{\max}. \quad (1.3)$$

In the analysis of measurement errors occurring in repeatable experiments, a division into systematic and random errors is made. Investigating the results of repeated measurements of the same quantity leads to the observation that one component of the error does not change its sign or value, or evolves with changes in the reference conditions according to a specific law (function). This component was named the systematic error or bias (Taylor 1997, Guide 1995). It is defined as follows: the **systematic error (bias)** is the difference between the mean value calculated for an infinite number of measurements of a quantity – carried out under the same conditions – and its actual value.

The second component of the error is commonly called the random error (Guide 2004). It can be reduced by repeating the measurement. In VIM (1993) the **random error** is defined as the difference between the result of the individual measurement and the mean value calculated for an infinite number of measurements of a quantity, carried out under the same conditions.

The above definitions of measurement errors refer to the results of individual measurements. When a measurement model is given in the form of a function of input quantities, it is called an indirect measurement. The error of an indirect measurement is determined on the basis of the law of error propagation (Taylor 1997). According to this law, to determine an output quantity error on the basis of known errors of input quantities, one of two methods can be used: the expansion of the model function in a Taylor series up to first-order terms (method of total differential); or the method of increments.

The **method of increments** (exact method) consists of determining the increment of a measurement model function for the known increments of input quantities (i.e. absolute errors). Let us consider a measurement model in the form of a function of several variables:

$$y = f(x_1, x_2, \ldots, x_n), \quad (1.4)$$

where $x_1, x_2, \ldots, x_n$ are inputs and y the measurement result. Let us also denote $\Delta x_1, \Delta x_2, \ldots, \Delta x_n$ as absolute errors of the inputs (increments of f arguments). We can then write the increment of the function as:

$$\Delta y = y + \Delta y - y. \quad (1.5)$$

The first two components on the right-hand side of (1.5) can be expressed as:

$$y + \Delta y = f(x_1 + \Delta x_1, x_2 + \Delta x_2, \ldots, x_n + \Delta x_n). \quad (1.6)$$

Finally:

$$\Delta y = f(x_1 + \Delta x_1, x_2 + \Delta x_2, \ldots, x_n + \Delta x_n) - f(x_1, x_2, \ldots, x_n). \tag{1.7}$$

Hence, from (1.7), the relative error of y is:

$$\delta y = \frac{f(x_1 + \Delta x_1, x_2 + \Delta x_2, \ldots, x_n + \Delta x_n) - f(x_1, x_2, \ldots, x_n)}{f(x_1, x_2, \ldots x_n)}. \tag{1.8}$$

The above method was used in this work in computer simulations of the method error in infrared thermography presented in Chapter 4.

Unfortunately, for complicated measurement models, evaluation of the error by means of (1.7) and (1.8) is very tedious. Therefore, the error is often evaluated by using an approximated method – the method of the total differential.

The **method of the total differential** (approximated method) is based on the expansion of the function $f(x_1, x_2, \ldots, x_n)$ as a Taylor series around the point defined by the actual (true conventional) values of the inputs. Assuming that function (1.4) is continuous and, for simplicity, that only input x_1 is burdened with error Δx_1, the expansion in a Taylor series has the following form:

$$f(x_1 + \Delta x_1, x_2, \ldots, x_n) = f(x_1, x_2, \ldots, x_n) + \frac{\Delta x_1}{1!} f'(x_1, x_2, \ldots, x_n)$$
$$+ \frac{(\Delta x_1)^2}{2!} f''(x_1, x_2, \ldots, x_n) + \frac{(\Delta x_1)^3}{3!} f'''(x_1, x_2, \ldots, x_n) + \ldots. \tag{1.9}$$

The terms of order higher than one can be omitted in the above expansion, assuming that their influence on the result is negligible. On the basis of (1.6), we can write:

$$\Delta y_1 = \Delta x_1 \cdot f'(x_1, x_2, \ldots, x_n) = \Delta x_1 \cdot \frac{\partial y}{\partial x_1}, \tag{1.10}$$

where Δy_1 is called the component of the output error associated with x_1. Partial derivative $\partial y / \partial x_1$, calculated at point $(x_1, x_2, \ldots, x_n)$, is called the sensitivity index to input x_1. If we take into account errors from all inputs $x_1, x_2 \ldots x_n$, the total error of the indirect measurement can be written as the sum:

$$\Delta y = \sum_{i=1}^{n} \Delta x_i \frac{\partial y}{\partial x_i}, \tag{1.11}$$

where partial derivatives $\partial y / \partial x_i$ are calculated at $(x_1, x_2, \ldots, x_n)$. Since in (1.11) all the increments of input variables $x_1, x_2, \ldots, x_n$ are taken with the same sign, the total error is overestimated. In real experiments, the probability that all input measurements are burdened with positive (or negative) errors is small and decreases with an increasing number of inputs (Fuller 1987). Therefore, a more realistic estimate of the indirect measurement absolute error is commonly used in practice – the mean square error:

$$\Delta y = \sqrt{\left(\Delta x_1 \frac{\partial y}{\partial x_1}\right)^2 + \left(\Delta x_2 \frac{\partial y}{\partial x_2}\right)^2 + \cdots + \left(\Delta x_n \frac{\partial y}{\partial x_n}\right)^2}. \tag{1.12}$$

From the point of view of temperature measurement using an infrared system, analysis of error can be useful only for strictly defined reference conditions. Such an analysis can also be helpful in the sensible estimation of measurement accuracy in situations where there is no information on these conditions. An additional aim of the analysis is the comparison of various measurement models used in contemporary infrared cameras. Lastly, analysis of error can be a starting point for the investigation of sensitivity, when thermography is applied to validate numerical models (e.g. when temperature measured at different points is used in finite element method (FEM) computations).

1.2 Uncertainties in Indirect Measurements

In accurate comparative measurements (such as standard measurements) it is necessary to describe the reference conditions in the form of random variables with assumed probability distributions. In such situations it is more convenient to use the concept of measurement uncertainty. In general, the measurement uncertainty characterizes the doubt about a measurement result. With this meaning, the uncertainty does not determine any specific quantitative measure. It only expresses the lack of accurate knowledge about a value of the measured quantity. Therefore, a measurement result is always an estimate of the measured quantity. More specifically, the measurement uncertainty is defined in the following way (VIM 1993): the **uncertainty of a measurement** is a parameter characterizing the spread of measurement values that can be assigned to the measured quantity in a justified way.

Unfortunately, the above definition does not determine how this assignment can be made. Therefore, to characterize the measurement accuracy precisely, the following definition of the standard uncertainty was introduced as a quantitative measure of spread (Guide 1995): the **standard uncertainty of a measurement** is the uncertainty of measurement values expressed in the form of the standard deviation.

To estimate the quantitative accuracy of a measurement, a description of the measurement model inputs in the form of random variables is introduced. These variables are characterized by specific probability distribution functions. For estimating the measurement accuracy, the most important statistics of a random variables are the expected value and the standard deviation.

The **expected value** $\boldsymbol{E(X)}$ of a discrete random variable X, whose values x_i appear with probabilities p_i, is:

$$E(X) = \sum p_i x_i, \tag{1.13}$$

where the sum is taken over all possible values x_i of variable X.

In practice, the set of measured values x_i is a finite N-element set. Therefore, the expected value is substituted by its estimator – the arithmetic mean from N independent observations (Söderström and Stoica 1994):

$$\bar{x} = \frac{1}{N}\sum_{i=1}^{N} x_i. \tag{1.14}$$

The **standard deviation** $\boldsymbol{\sigma(X)}$ of a random variable is the positive square root of the variance:

$$\sigma(X) = \sqrt{E[X - E(X)]^2}. \tag{1.15}$$

In practical problems an estimator of the standard deviation, called the experimental standard deviation, is used. It is calculated from N independent observations x_i:

$$s(x_i) = \sqrt{\frac{1}{N-1}\sum (x_i - \bar{x})^2}. \tag{1.16}$$

In accordance with Recommendation INC-1 (1980), the components of a measurement uncertainty can be grouped into two categories (CIPM 1981, CIPM 1986):

Type A standard uncertainty determined on the basis of the observed frequency distribution.
Type B standard uncertainty determined on the basis of a frequency distribution assumed a priori.

Example 1.1 *Estimation of the parameters of a probability density function from a series of measurements – Type A uncertainty*

In this example we simulated an experiment consisting of multiple measurements of quantity X under repeated conditions. The uncertainty analysis was conducted on the basis of a series of realizations generated to simulate the results of real measurements. The estimation of parameters of the density function was conducted assuming that the measured series was subject to a Gaussian distribution (this distribution was determined on the basis of the shape of the histogram). The results of numerical calculations are shown in Figure 1.1a. The solid line denotes the probability density function obtained for the estimated parameters. To evaluate the standard uncertainty, we used the arithmetic mean (1.14) and the experimental standard deviation 1.16 as the best estimators of the expected value and the standard uncertainty respectively. Hence, we could use MATLAB's functions `mean()` and `std()` to determine these basic statistics of the obtained distribution. This experiment illustrates how to evaluate Type A standard uncertainty.

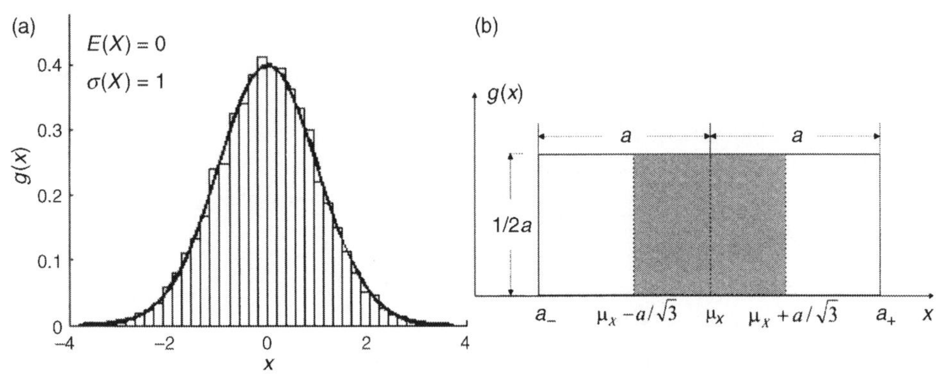

Figure 1.1 Evaluation of standard uncertainty of: (a) Type A, simulation in MATLAB; and (b) Type B, for uniform probability distribution (Guide 1995)

Example 1.2 *Evaluation of Type B standard uncertainty from the parameters of a uniform distribution density function*

Figure 1.1b shows how Type B standard uncertainty can be determined assuming the uniform probability distribution of variable X. The distribution density function is:

$$\begin{aligned} g(x) &= 1/2a \quad \text{for} \quad a_- \leq x \leq a_+ \\ g(x) &= 0 \qquad\; \text{for other } x. \end{aligned} \tag{1.17}$$

In this example, to determine the Type B standard uncertainty, we used information on the permissible interval of the measured values. The assumption of a uniform distribution is the worst case, for which the standard uncertainty is $a/\sqrt{3}$, where a is one-half of the interval length.

In the case of errors, the problem on how individual standard uncertainties of inputs of a complex analytical model affect the precision of an indirect measurement (i.e. evaluation of a measurement uncertainty) often occurs in practice. In such a situation it is necessary to evaluate the combined standard uncertainty. Depending on whether the model inputs are correlated or not, a covariance factor appears in the definition of the combined uncertainty. Provided that the input variables are uncorrelated, the combined standard uncertainty is defined according to VIM (1993): the **combined standard uncertainty** $\boldsymbol{u_c(y)}$ is the positive square root of the combined variance $u_c^2(y)$, defined as:

$$u_c^2(y) = \sum_{i=1}^{N} \left(\frac{\partial f}{\partial x_i}\right)^2 u^2(x_i), \tag{1.18}$$

where $y = f(x_1, x_2, \ldots, x_n)$ is the measurement model function (1.4) and $u^2(x_i)$ is the variance of the ith input of the model.

When the input variables are correlated, the expression describing the uncertainty is more complicated because it includes estimates of the covariance of the inputs. The combined uncertainty of measurement y is determined as (Taylor 1997):

$$u_c^2(y) = \sum \left(\frac{\partial f}{\partial x_i}\right)^2 u^2(x_i) + 2\sum_{i=1}^{n-1} \sum_{j=i+1}^{n} \frac{\partial f}{\partial x_i}\frac{\partial f}{\partial x_j} u(x_i, x_j), \tag{1.19}$$

where $u(x_i, x_j)$ is the estimate of covariance between x_i and x_j.

Because in the evaluation of indirect measurement uncertainty the inputs of the model are considered as random variables, the determined estimators (expected values, standard deviations) are also random variables. That is why we need to define the determined parameters using concepts of probability. These concepts are (VIM 1993) discussed below.

The **one-sided coverage interval** is as follows. If T is a function of observed values, such that for estimated parameter of population θ, probability $Pr(T \geq \theta)$ or $Pr(T \leq \theta)$ is at least equal to $(1 - \alpha)$ (where $(1 - \alpha)$ is a fixed number, positive and smaller than one), then the interval from the smallest possible value of θ to T (or the interval from T to the biggest possible value of θ) is the one-sided coverage interval θ with confidence level $(1 - \alpha)$.

The **confidence level** is the value $(1 - \alpha)$ of probability associated with a confidence interval or statistical coverage interval.

Estimation of the combined standard uncertainty is usually associated with the simultaneous evaluation of probability, with which a measurement result lies inside the interval determined by this uncertainty. For the strict determination of this probability, the concept of so-called expanded uncertainty is introduced: the **expanded uncertainty** U is the uncertainty obtained by multiplying the combined standard uncertainty $u_c(y)$ by expansion factor k:

$$U = ku_c(y). \tag{1.20}$$

The expanded uncertainty specifies the limits of the uncertainty interval for a given confidence level. The value of the expansion factor depends on the probability distribution of the model output variable. For example, if a random variable has a Gaussian distribution at the model output, the probability that a measurement result will fall in the interval from $y - u_c(y)$ to $y + u_c(y)$, that is for $k = 1$, is about 68%; from $y - 2u_c(y)$ to $y + 2u_c(y)$, that is for $k = 2$, about 95%; and from $y - 3u_c(y)$ to $y + 3u_c(y)$, that is for $k = 3$, about 99%. The requirement for exact knowledge of the type of the output variable distribution is an inconvenience in determining the expansion factor. For models with a large number of inputs it can be assumed that the central limit theorem applies. In such a case it is assumed a priori that the output variable has a Gaussian distribution. However, measurement practice reveals significant differences in evaluated expanded uncertainties obtained under the assumption of the Gaussian distribution, especially when the measurement model exhibits strong nonlinearities and the true distribution of the output variable is asymmetric.

When the probability distribution of the model output variable is not known, we need to determine the relationship between the expansion factor and the coverage interval (confidence level). This problem is usually solved by determining the resultant number of degrees of freedom ν from the Welch–Satterthwaite equation (Welch 1936, Satterthwaite 1941]:

$$\nu_{eff} = \frac{u_c^4(y)}{\sum_{i=1}^{N} \frac{u_i^4(y)}{\nu_i}}, \tag{1.21}$$

and calculating the expanded uncertainty as (Guide 1995):

$$U_p = k_p u_c(y) = t_p(\nu_{eff}) u_c(y), \tag{1.22}$$

where coefficient $t_p(\nu_{eff})$ is the value of Student's t distribution calculated on the basis of the number of degrees of freedom approximated by formula (1.21).

Summing up, the steps in evaluating the combined standard uncertainty can be presented as follows (Guide 2004):

1. Evaluation of the expected values and standard deviations $u(x) = (u(x_1)\ldots u(x_n))^T$ of probability distributions of random variables $X_1 \ldots X_n$ representing inputs $x = (x_1 \ldots x_n)^T$ of the measurement model. If the inputs are correlated with each other, we should use joint probability distributions of the variables.
2. Evaluation of covariances (reciprocal uncertainties) $u(x_i, x_j)$ as $Cov(X_i, X_j)$.
3. Evaluation of partial derivatives of measurement model (1.4) with respect to its inputs.
4. Calculation of estimate of the output y on the basis of the measurement model function f.
5. Evaluation of sensitivity indices of the model using the partial derivatives determined in step 3, calculated at point $\mathbf{x} = (x_1 \ldots x_n)$.
6. Evaluation of combined standard uncertainty $u_c(y)$ on the basis $u(\mathbf{x})$, $u(x_i, x_j)$ and the model sensitivity indices, using formula 1.18 or 1.19.

7. Evaluation of the number of degrees of freedom using, for example, formula (1.21).
8. Calculation of the expanded uncertainty from (1.22).

These steps lead to correct evaluation of the expanded uncertainty only when the three following conditions are satisfied:

1. Nonlinearity of the model is negligible. Because it is difficult to indicate an objective measure of nonlinearity, we mean here that neglecting the higher order terms in the Taylor series expansion of the model function does not greatly affect the results of the estimation.
2. The assumptions of the central limit theorem are satisfied. In particular, the distribution of the model output is Gaussian with good accuracy.
3. The approximation of the resultant number of degrees of freedom from the Welch–Satterthwaite equation is accurate enough.

In practical measurements it is often not possible to fulfill the above conditions. Consequently, Working Group No. 1 of the Common Committee for Basic Problems in Metrology produced Supplement No. 1 to Guide (1995) entitled 'Numerical methods for the propagation of distributions'. In this supplement, an idea of the numerical evaluation of the coverage interval is presented. With such an approach, knowledge of an analytical form of the probability distribution function is not required. In further parts of this book we present the basic guidelines and aims of this method.

1.3 Method for the Propagation of Distributions

In the previous points we presented the basic concepts associated with the evaluation of precision in indirect measurements making use of complex mathematical models. We described the basics of the theory of errors and uncertainties and indicated the problems arising in the evaluation of the expanded uncertainty. These problems are consequences of the fact that in practice the types of probability distributions of input random variables (i.e. measured quantities) are unknown.

The Common Committee for Basic Problems in Metrology took this into consideration when producing the supplement mentioned immediately above (Guide 2004). This supplement deals with the evaluation of precision in indirect measurements, with particular emphasis on strongly nonlinear and/or complicated measurement models, such as the processing algorithm of an infrared camera measurement path. The propagation of distributions method allows for the correct estimation of measurement precision, in particular in the following cases (Dudzik and Minkina 2007):

- partial derivatives are unavailable;
- the distribution of the output variable is not Gaussian;
- the distributions of the input variables exhibit asymmetry;
- the measurement model is a strongly nonlinear function of input quantities;
- uncertainty ranges of individual input quantities are incomparable.

The idea of the propagation of distributions is illustrated in Figure 1.2.

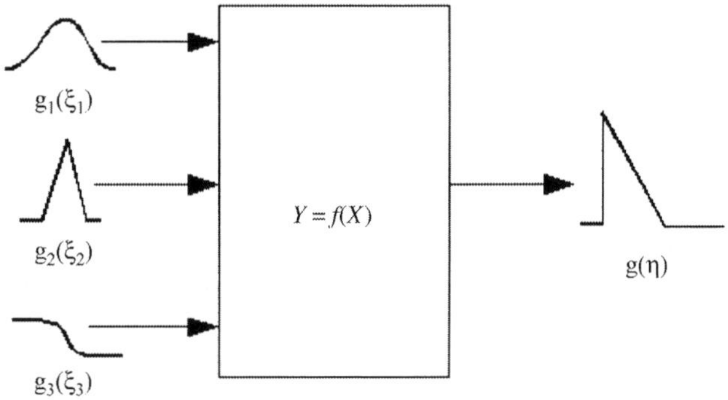

Figure 1.2 Illustration of the propagation of distributions

The symbols in Figure 1.2 denote: $g_i(\xi_i)$, the probability density functions of permissible values ξ_i of the ith input quantity X_i; and $g(\eta)$, the probability density function of permissible values η of output quantity Y of measurement model $Y = f(X)$.

In the method for the propagation of distributions the uncertainties are evaluated using the Monte Carlo method. The principal aim of the computational procedure is to evaluate the statistical coverage interval at a specified confidence level. It is worth emphasizing that the procedure gives correct results even for strongly nonlinear functional relationships of measurement models as well as for asymmetric probability density functions of input random variables. The following steps can be distinguished in the evaluation of the uncertainty:

1. Definition of the output quantity of the considered measurement model (indirectly measured quantity).
2. Definition of input quantities of the model.
3. Design of the measurement model on the basis of available (experimental or theoretical) knowledge of the measured quantity.
4. Determination of shapes of probability density functions of the model inputs (based on analysis of a series of input measurements or on a single experiment).
5. Evaluation of the probability distribution of the output using the measurement model and the determined distributions of the inputs. The calculations can be carried out using the Monte Carlo method.
6. Estimation of parameters of the resulting probability density function; that is, the standard uncertainty and the corresponding expected value of the output as well as the coverage (confidence) interval that includes the measurement results with probability determined by an assumed confidence level.

The Monte Carlo method makes possible the numerical approximation of cumulative distribution $G(\eta)$ of the output quantity. The simulation is based on the assumption that any value of an input quantity chosen at random from all permissible values of this input is as justified as any other. In other words, no value is preferred. Hence, drawing values of each input quantity according to the probability distribution function assigned to this input validates the set of its values. The value of the measurement model output corresponding to the drawn values of the inputs is a representative output. Consequently, a big enough set of output values obtained from the model in this way can approximate, to the required accuracy, the probability density distribution of permissible values of the output (measured quantity). The Monte Carlo simulation is performed in the following steps (Guide 2004):

1. Generation of a set of N values by independent sampling of the probability density function of each input variable $X_i, i = 1, \ldots, N$. In the case of statistically dependent variables the samples must be generated with the use of the joint density function of the variables. The sampling is repeated M times, where M is a large number. As a result, we obtain M independent sets of N values of the inputs.
2. Simulation of the model for each set of values. As a result, we obtain a set of M values (realizations) of the model output variable Y. This set is in fact a numerical approximation of the probability density distribution of the output variable.
3. Determination of approximation $\hat{G}(\eta)$ of cumulative density function $G(\eta)$ of Y, based on the generated set of values.
4. Evaluation of statistical parameters of the output variable distribution on the basis of $\hat{G}(\eta)$. In particular, this determines: the measured value y as the expected value of $\hat{G}(\eta)$; the estimate of standard uncertainty $u(y)$ as the standard deviation of $\hat{G}(\eta)$; and the end points of coverage interval $I_p(y)$ for the assumed coverage probability, as two quantiles of $\hat{G}(\eta)$.

One very important aspect of the propagation of distributions is the approximation of the cumulative density function of the output quantity. The steps for the approximation procedure are as follows:

1. Sorting of values $y_r, r = 1, \ldots, M$, of the output variable (obtained from the Monte Carlo simulation) in non-decreasing order. The sorted values are further denoted as $y_{(r)}, r = 1, \ldots, M$.
2. Assignment of equidistant cumulative probabilities to the sorted values according to the formula (Cox *et al.* 2001):

$$p_r = \frac{r - 0.5}{M}, \quad r = 1, \ldots, M. \tag{1.23}$$

3. Forming of piecewise linear function $\hat{G}(\eta)$ by joining M points of coordinates $(y_{(r)}, p_r)$:

$$\hat{G}(\eta) = p_r + \frac{\eta - y_{(r)}}{M(y_{(r+1)} - y_{(r)})}, \quad y_{(r)} \le \eta \le y_{(r+1)}, \quad r = 1, \ldots, M - 1. \tag{1.24}$$

Having determined approximation $\hat{G}(\eta)$ of the output probability distribution, it is possible to calculate its expected value $\hat{y}$, which is an estimate of measured quantity Y, and its standard deviation, which is an estimate of the standard uncertainty. Estimates of the expected value and the variance can be calculated as:

$$\hat{y} = \frac{1}{M}\sum_{r=1}^{M} y_r \tag{1.25}$$

and:

$$u_c^2(\hat{y}) = \frac{1}{M-1}\sum_{r=1}^{M}(y_r - \hat{y})^2. \tag{1.26}$$

The last step in the propagation of distributions algorithm is the evaluation of the coverage interval resulting from the assumed confidence level (coverage probability). Usually, the 95% confidence level is adopted.

The **quantile of order $\boldsymbol{\beta}$** of a probability distribution described by cumulative distribution function $G(\eta)$ is such that a value η of the random variable satisfies the equality $G(\eta) = \beta$. This means that the probability of occurrence of this value is equal to β.

If we denote by α a value for the interval from 0 to $1-p$, where p is the required coverage probability, then the ends of the coverage interval $I_p(y)$ can be determined as quantiles of order α and $\alpha + p$ of the distribution defined by $G(\eta)$. For example, if we adopt $\alpha = 0.025$, the ends of the 95% coverage interval will be quantiles of order 0.025 and 0.0975. As a result, we obtain the probabilistically symmetric coverage interval $I_{0.95}(y)$.

In general, if a probability distribution is symmetric, the shortest coverage interval is associated with quantile:

$$\alpha = \frac{1-p}{2}. \tag{1.27}$$

As we can see, for the 95% coverage interval and provided that the distribution is symmetric and satisfies (1.27), we obtain $\alpha = 0.025$; that is, the value used in the above example.

The Monte Carlo simulation may reveal that the probability distribution density of the output is not symmetric with respect to its expected value (it is not the centered expected value). In this case there are many intervals satisfying equality:

$$g(G^{-1}(\alpha)) = g(G^{-1}(\alpha+p)), \tag{1.28}$$

and we should choose such a value of α that determines the shortest possible coverage interval associated with assumed probability p. The value of α chosen in this way satisfies the condition:

$$\hat{G}^{-1}(\alpha+p) - \hat{G}^{-1}(\alpha) = \min. \tag{1.29}$$

Below we present an example of uncertainty analysis using the propagation of distributions. (The procedure described in this example corresponds in principle to a much more complicated case study: the simulation evaluation of uncertainty of an infrared camera processing path algorithm. The methodology and results of such an evaluation are presented in Chapter 5.)

Example 1.3 *Application of the propagation of distributions for the determination of the 95% coverage interval of a simple nonlinear model*

Let us consider a simple measurement model with two input and one output random variables. The relationship between the inputs and the output is:

$$Y = X_1^2 + 2X_2. \tag{1.30}$$

Table 1.1 Input data for the Monte Carlo simulation of model (1.30)

Model input quantity	$\hat{x}$	$u^2(x)$	a	b
X_1	1.0	2.0	7.550	12.45
X_2	100	200	92.25	107.7

Input variables X_1 and X_2 are subject to uniform probability distributions defined as:

$$g(x) = \begin{cases} \dfrac{1}{b-a} & \text{for } a \leq x \leq b \\ 0 & \text{for other } x. \end{cases} \tag{1.31}$$

Parameters a and b of these distributions can be calculated from given statistics of the inputs, namely the expected values and the variances (squared standard deviations):

$$\begin{aligned} a &= \hat{x} - \sqrt{3u^2(x)} \\ b &= \hat{x} + \sqrt{3u^2(x)}. \end{aligned} \tag{1.32}$$

In the above equations we assume that the expected value is also an estimate of the input, denoted as $\hat{x}$, and $u^2(x)$, denoted as an estimate of the input variance. The adopted values of the estimates and uncertainties for inputs X_1 and X_2, as well as the corresponding parameters of the uniform distribution functions calculated from these estimates, are shown in Table 1.1. They are the input data for the Monte Carlo simulation.

The Monte Carlo simulation consisted of $M = 10^5$ runs and generated approximate distributions of the input variables X_1 and X_2 for the conditions determined by the parameters from Table 1.1. These distributions are presented in Figures 1.3 and 1.4.

In order to determine the probability density function of output variable Y we also performed M runs of the Monte Carlo simulation of model (1.30). The approximation of the cumulative

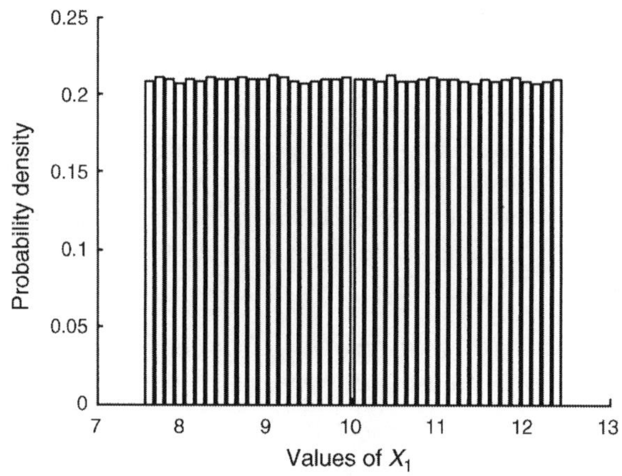

Figure 1.3 Probability density distribution of input variable X_1

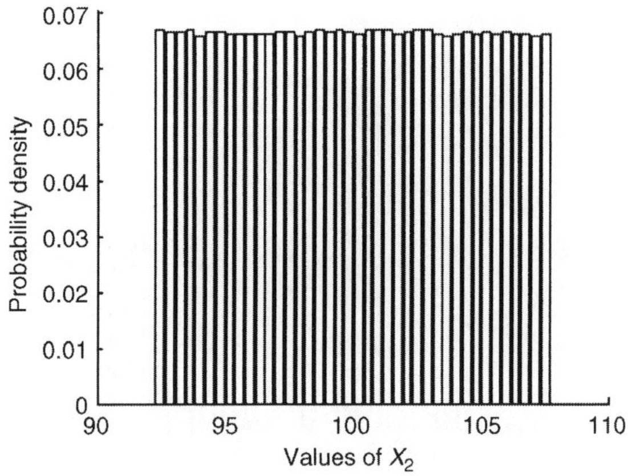

Figure 1.4 Probability density distribution of input variable X_2

distribution function of Y was determined using formulas (1.23) and (1.24). This approximation of the cumulative distribution is shown in Figure 1.5, and the corresponding probability density approximation is presented in Figure 1.6. The estimate of the measured quantity, calculated as the arithmetic mean of variable Y, is $\hat{y} = 302$, standard uncertainty $u_c(\hat{y}) = 30$, and the 95% coverage interval, marked by vertical lines in Figure 1.6, is $I_{0.95}(Y) = [252, 354]$.

The above example explains the propagation of distributions for a simple measurement model: a second-order polynomial. In a further part of this monograph we discuss the application of this method to much more complex models of measurement in infrared thermography (described in detail in Chapter 3). For such a complex model, the use of the propagation of distributions is justified more than an analytic approach. In addition, it yields

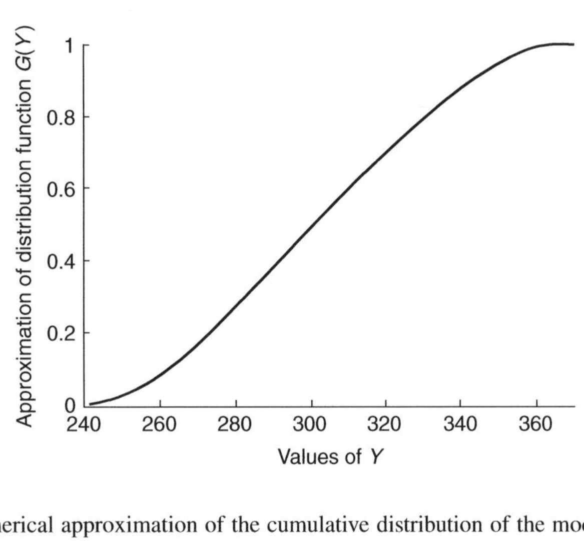

Figure 1.5 Numerical approximation of the cumulative distribution of the model output variable

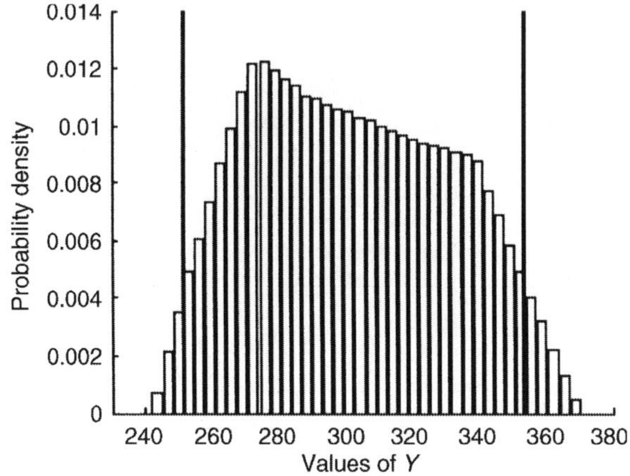

Figure 1.6 Probability density function with marked 95% shortest coverage interval

more accurate results due to the model's nonlinearity. The methodology of the simulation and investigation of the infrared thermography measurement model is presented in Chapter 5.

Summing up the above considerations on the basic concepts of metrology, we want to emphasize that error analysis and uncertainty analysis do not exclude each other. One of our aims is to present the analysis of errors and uncertainties as complementary methods of evaluating accuracy in infrared thermography measurements.

2

Measurements in Infrared Thermography

2.1 Introduction

F.W. Herschel's experiment, resulting in the discovery of infrared radiation, had fundamental significance for the rise and development of research in the infrared spectrum. In his experiment Herschel (1800a–d, 1830) observed heat effects associated with different spectral ranges of the Sun's radiation. He placed blackened containers of sensitive mercury thermometers along the spectrum obtained by splitting the Sun's radiation in a glass prism. The energy of incident rays was absorbed by the containers and the thermometers indicated temperatures higher than the ambient temperature. On investigating the results of his experiment, Herschel discovered that the readings of the thermometers located beyond the red end of the spectrum were higher than those of the thermometers located within the visible range. The experiment proved that the full spectrum of the Sun's radiation is wider than the visible range, and that directly beyond the red end there exist rays, weakly refracted by the prism, invisible to the naked eye.

Herschel named them 'invisible rays' or the 'invisible thermometrical spectrum'. Somewhat later they were commonly named 'infrared rays'. Further experiments carried out by other scientists showed that the infrared rays – likewise visible rays – are subject to reflection, refraction and absorption. Experiments on interference and polarization proved that infrared radiation has the same nature as visible radiation. However, it was Herschel who first determined the point of maximum heat effect and stated that it is located beyond the visible range of the spectrum.

Although Herschel is considered as a pioneer in infrared research, it is difficult to imagine contemporary infrared thermography without the basic laws of radiative heat transfer. They are formulated and described below.

2.2 Basic Laws of Radiative Heat Transfer

At temperatures higher than 0 K, the absolute zero temperature, each body emits heat (thermal) radiation. The intensity of this radiation depends on wavelength λ and the body's temperature (Gerashenko *et al.* 1989).

Infrared Thermography: Errors and Uncertainties Waldemar Minkina and Sebastian Dudzik

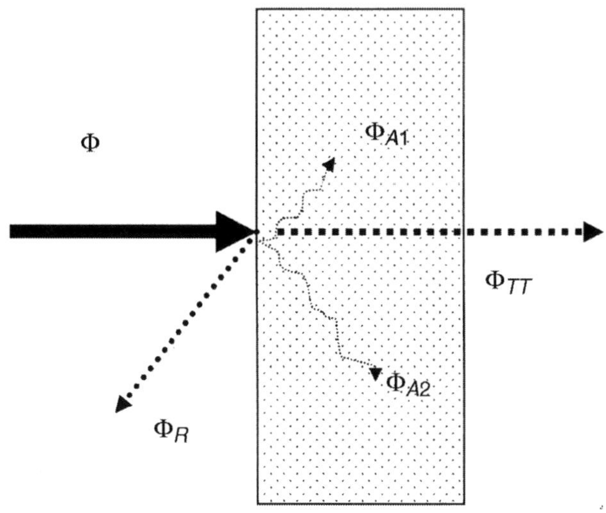

Figure 2.1 Paths of fluxes of thermal radiation incident on a body of a certain thickness

Thermal radiation is a kind of electromagnetic radiation commonly occurring in nature. If heat flux Φ (amount of heat in a time unit) (W) falls on the surface of a body of a certain thickness, and flux Φ_A (W) is absorbed, flux Φ_R (W) reflected and flux Φ_{TT} (W) transmitted across the body, then the following coefficients are introduced: absorbance coefficient A, reflectance coefficient R and transmission coefficient TT. They are defined as ratios:

$$A = \frac{\Phi_A}{\Phi}, \qquad R = \frac{\Phi_R}{\Phi}, \qquad TT = \frac{\Phi_{TT}}{\Phi}. \tag{2.1}$$

Schematic paths of these three heat fluxes are sketched in Figure 2.1.

The concept of a perfect black body plays a very important role in infrared thermography measurements. Several models of black bodies are shown in Figure 2.2 (McGee 1998, Ricolfi and Barber 1990). A black body completely absorbs incident radiation so the coefficients defined by (2.1) satisfy equalities:

$$A = 1, \qquad R = 0, \qquad TT = 0. \tag{2.2}$$

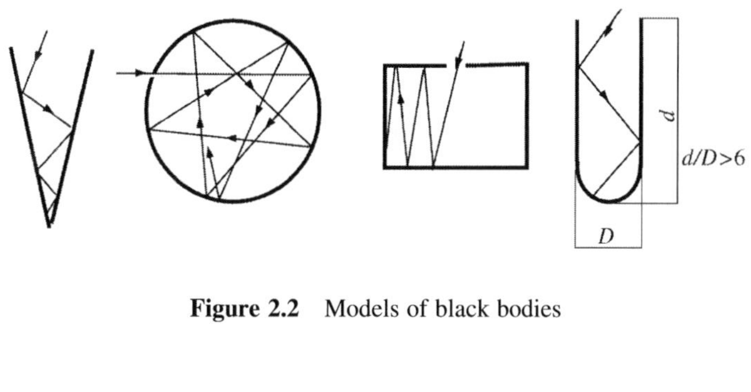

Figure 2.2 Models of black bodies

Perfect absorption of incident radiation in the above models results from multiple inner reflections.

The following equality – Kirchhoff's law – is satisfied for each body:

$$A + R + TT = 1. \tag{2.3}$$

This law is fulfilled not only for the overall radiation (emitted by a body over the full range of wavelengths), but also for any monochromatic radiation. For monochromatic radiation, spectral absorptance A_λ, reflectance R_λ and transmission TT_λ are introduced. Their values depend on specified wavelength λ and are defined as (Janna 2000, Bayazitoğlu and Özişik 1988, Wiecek *et al.* 1998, Wiecek 1999, Wolfe and Zissis 1978):

$$A_\lambda = \frac{\Phi_{\lambda A}}{\Phi_\lambda}, \qquad R_\lambda = \frac{\Phi_{\lambda R}}{\Phi}, \qquad TT_\lambda = \frac{\Phi_{\lambda TT}}{\Phi_\lambda}. \tag{2.4}$$

Kirchhoff's law applies also for the spectral coefficients:

$$A_\lambda + R_\lambda + TT_\lambda = 1. \tag{2.5}$$

The values of coefficients A, R and TT depend on the material of a body and its surface state, whereas the spectral coefficients depend also on the wavelength. It should be emphasized that, in general, these coefficients depend also on temperature T. Some authors (e.g. Madura *et al.* 2001) have announced that for ultrafast thermal processes the coefficients also depend on time τ. Below we present basic definitions associated with the radiative heat transfer.

The ratio of (temperature- and wavelength-dependent) radiant power (radiant flux) $d\Phi$, emitted by an arbitrarily small element of surface containing the considered point, to projected area dF of that element, is called **radiant exitance (emittance)**, and the radiant flux per unit solid angle is called **radiant intensity** and denoted I (W sr^{-1}). Radiant exitance is expressed as (Kreith 2000, Hudson 1969):

$$M(\lambda, T) = \frac{d\Phi(\lambda, T)}{dF}, \mathrm{W\,m^{-2}}. \tag{2.6}$$

Thermal flux density q is expressed in the same units as the radiant exitance:

$$q(\lambda, T) = \frac{d\Phi(\lambda, T)}{dF}, \mathrm{W\,m^{-2}}. \tag{2.7}$$

Monochromatic radiant exitance is defined as:

$$M_\lambda(\lambda, T) = \frac{dM(\lambda, T)}{d\lambda}, \mathrm{W\,m^{-2}\,\mu m^{-1}}. \tag{2.8}$$

Spectral radiant exitance for a black body is given by Planck's law (Özisik 1973):

$$M_b(\lambda, T) = \frac{2\,\pi\,h\,c^2}{\lambda^5\left[\exp\left(\frac{h\,c}{\lambda\,k\,T}\right) - 1\right]}, \mathrm{W\,m^{-2}\,\mu m^{-1}}, \tag{2.9}$$

where (following Wark (1988) and ISO 31): $c = 299\,792\,458 \pm 1.2\,\mathrm{m\,s^{-1}}$, the speed of light in vacuum; $h = (6.626\,176 \pm 0.000\,036)\cdot 10^{-34}\,\mathrm{W\,s^2}$, the Planck constant; and $k = (1.380\,662 \pm 0.000\,044)\cdot 10^{-23}\,\mathrm{W\,s\,K^{-1}}$, the Boltzmann constant.

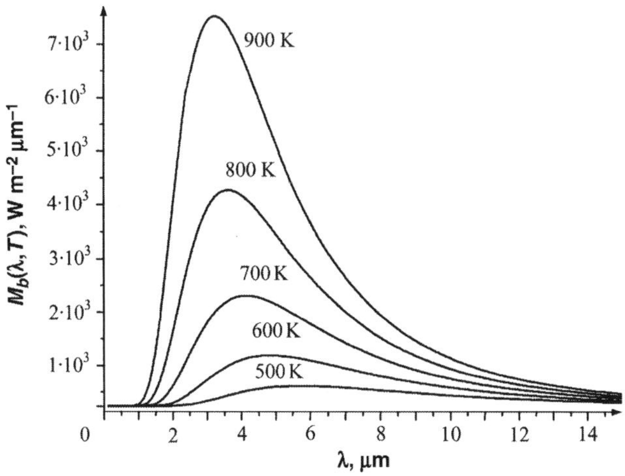

Figure 2.3 Radiant exitance $M_b(\lambda,T)$ of a black body according to Planck's formula (2.10) (Minkina 2004)

By defining new constants $c_1 = 2\cdot\pi\cdot h\cdot c^2 = (3.741\ 832 \pm 0.000\ 020)\cdot 10^{-16}\,\mathrm{W\,m^2}$ (the so-called first radiant constant), $c_2 = h\cdot c/k = (1.438\,786 \pm 0.000\,045)\cdot 10^{-2}\,\mathrm{m\,K}$ (the so-called second radiant constant) according to the International Temperature Scale, ITS-90, formula (2.9) can be written in the more compact form:

$$M_b(\lambda, T) = \frac{c_1}{\lambda^5 \cdot \left[\exp\left(\frac{c_2}{\lambda \cdot T}\right) - 1\right]}, \mathrm{W\,m^{-2}\,\mu m^{-1}}. \tag{2.10}$$

The graphs in Figure 2.3 show radiant exitance $M_b(\lambda,T)$ of a black body, determined by (2.10), versus wavelength λ for different temperatures T (Minkina 2004).

The **band radiant exitance** $\boldsymbol{M_b(\lambda_1,\lambda_2)}$ of a black body is the integral of spectral intensity (2.10) over a band from wavelength λ_1 to wavelength λ_2:

$$M_b(\lambda_1 \div \lambda_2) = \int_{\lambda_1}^{\lambda_2} \frac{c_1 d\lambda}{\lambda^5 \cdot \left[\exp\left(\frac{c_2}{\lambda \cdot T}\right) - 1\right]}. \tag{2.11}$$

Planck's law determines the radiant exitance $M_b(\lambda,T)$ of a black body for given temperature T and wavelength λ. Sometimes there is a need to determine a black body temperature T for $M_b(\lambda)$ measured for a certain λ (Chrzanowski 2000). This can be done from the inverse Planck's law:

$$T = \frac{c_2}{\ln\left[\frac{c_1 + \lambda^5 \cdot M_b(\lambda)}{\lambda^5 \cdot M_b(\lambda)}\right]^{\lambda}}, \mathrm{K}. \tag{2.12}$$

In certain situations Planck's law can be simplified. These particular situations are described by Wien's law and the Rayleigh–Jeans law.

Wien's law is an approximation of Planck's law for small values of product λT. In such a situation, the approximation $\exp(c_2/\lambda T) - 1 \approx \exp(c_2/\lambda T)$ holds. For the radiant exitance of a black body, Wien's law is expressed by:

$$M_b(\lambda, T) = \frac{c_1}{\lambda^5 \cdot \exp\left(\frac{c_2}{\lambda \cdot T}\right)}, \mathrm{W\,m^{-2}\,\mu m^{-1}} \tag{2.13}$$

The relative error due to substituting formula (2.13) for (2.10) is (Minkina 2004):

$$\delta = \frac{M_b(\lambda, T)_P - M_b(\lambda, T)_W}{M_b(\lambda, T)_P} = \exp\left(-\frac{c_2}{\lambda \cdot T}\right), \tag{2.14}$$

where $M_b(\lambda,T)_P$ and $M_b(\lambda,T)_W$ are radiant exitances calculated from Planck's law (2.10) and Wien's law (2.13) respectively.

The **Rayleigh–Jeans law** is an approximation of Planck's law for $\lambda T \gg c_2$. In such a situation the denominator of formula (2.10) is expanded into series:

$$\exp\left(\frac{c_2}{\lambda \cdot T}\right) - 1 \approx \frac{c_2}{\lambda \cdot T} + \frac{1}{2!}\left(\frac{c_2}{\lambda \cdot T}\right)^2 + \ldots, \tag{2.15}$$

and the higher order terms of this expansion are neglected. It leads to the following Rayleigh–Jeans formula for the radiant exitance of a black body:

$$M_b(\lambda, T) = \frac{c_1}{c_2} \cdot T \cdot \lambda^{-4}, \mathrm{W\,m^{-2}\,\mu m^{-1}}. \tag{2.16}$$

The relative error due to substituting formula (2.16) for (2.10) is:

$$\delta = 1 - \frac{T \cdot \lambda}{c_2}\left[\exp\left(\frac{c_2}{\lambda \cdot T}\right) - 1\right]. \tag{2.17}$$

Wien's displacement law is derived by equating to zero the derivative of function (2.10) with respect to wavelength λ:

$$\frac{dM_b(\lambda, T)}{d\lambda} = \frac{d}{d\lambda}\left\{\frac{c_1}{\lambda^5 \cdot \left[\exp\left(\frac{c_2}{\lambda \cdot T}\right) - 1\right]}\right\} = 0. \tag{2.18}$$

This equation determines wavelength $\lambda_{\max}$, for which the radiant exitance of a black body at a given temperature T reaches a maximum:

$$\lambda_{\max} \cdot T = 2898\,\mathrm{\mu m\,K}. \tag{2.19}$$

The maximum radiant exitance predicted by Wien's displacement law is:

$$M_b(T) = 1.286 \cdot 10^{-11} \cdot T^5, \mathrm{W\,m^{-2}\,\mu m^{-1}}. \tag{2.20}$$

The **Stefan–Boltzmann law** determines the total exitance for a black body at all wavelengths. This total exitance is obtained by integrating formula (2.10) from zero to infinity:

$$M_b = \int_{\lambda=0}^{\lambda=\infty} M_b(\lambda, T) d\lambda = \int_{\lambda=0}^{\lambda=\infty} \frac{c_1 d\lambda}{\lambda^5 \cdot \left[\exp\left(\frac{c_2}{\lambda \cdot T}\right) - 1\right]}. \tag{2.21}$$

The final Stefan–Boltzmann formula has the following form:

$$M_b(\lambda, T) = \frac{\pi^4}{15}\frac{c_1}{c_2^4} \cdot T^4 = \sigma_o \cdot T^4 = C_o\left(\frac{T}{100}\right)^4, \text{W m}^{-2}, \tag{2.22}$$

where:

$$\sigma_o = \frac{\pi^4}{15}\frac{c_1}{c_2^4} = \frac{2 \cdot \pi^5 \cdot k^4}{15 \cdot h^3 \cdot c^2} = (5.670\,32 \pm 0.000\,71) \cdot 10^{-8}\,\text{W m}^{-2}\,\text{K}^{-4}$$

is the Stefan–Boltzmann constant, and $C_o = \sigma_o \cdot 10^8 = (5.670\ 32 \pm 0.000\ 71)\,\text{W m}^{-2}\,\text{K}^{-4}$ is the technical constant of black body radiation.

Example 2.1 *Calculate an approximate value for the temperature of the Sun's surface. For the calculations it should be assumed that the maximum wavelength of the Sun's radiation, λ_{max}, is (approximately) one-half of the visible range (i.e. $\lambda_{max} \approx 0.50\,\mu m$)*

Using Wien's displacement law:

$$T = 2898/\lambda_{\max} = 2898/0.50 \approx 5800\,\text{K}.$$

It is necessary to emphasize that the method of calculation of the Sun's surface temperature is approximate because the exact value is not known. The exact value of this temperature can be determined by measuring the spectral radiant exitance $M_b(\lambda,T)$ of the Sun and applying the Stefan–Boltzmann law (2.22).

Example 2.2 *Calculate how much heat radiates from the human skin over a surface of $1\,m^2$ and a temperature of 310 K (36.9 °C)*

On the basis of the Stefan–Boltzmann law:

$$M_b = \sigma_o \cdot T^4 = 5.67 \cdot 10^{-8} \cdot (310)^4 \approx 500\,\text{W m}^{-2}.$$

All the above laws and definitions refer to black bodies. Unfortunately, a black body can only be considered as an idealized model of a real body. In reality, objects of infrared thermography measurements are not perfect absorbers of incident radiation – they are gray bodies. Therefore, one very important concept for an explanation of the operation of contemporary infrared systems, as well as for the correct recognition of error and uncertainty sources in measurements in infrared thermography, is the concept of emissivity. It will be considered in the next section.

2.3 Emissivity

2.3.1 Basic Concepts

The most important feature of a surface that affects the amount of energy radiating from it in stationary thermal conditions (fixed temperature) is its emissivity. If a surface whose temperature is to be measured had the properties of a black body, the radiant exitance for fixed temperature and wavelength could be determined from Planck's law (2.10). However, under real conditions Planck's law determines only a limiting (maximum) estimate of the thermal flux density. This is a consequence of the fact that all physical bodies have limited absorbing capacity; that is, they do not satisfy Planck's postulate referring to a black body (perfect black body). Therefore, it is necessary to introduce a parameter determining the absorbing capacity of a body's surface. On the basis of Kirchhoff's law, this is equivalent to determining the emissivity of a considered surface (Kreith 2000, Gaussorgues 1994, Glückert 1992).

The **emissivity** ε of a body over the full radiation range, called the total emissivity, is the ratio of full-range radiant exitance $M(T)$ of that body to full-range radiant exitance $M_b(T)$ of a black body at the same temperature:

$$\varepsilon_\lambda = \frac{M(T)}{M_b(T)}. \tag{2.23}$$

Monochromatic emissivity ε_λ is the ratio of monochromatic radiant exitance $M_\lambda\,(\lambda,T)$ of a body at a given wavelength λ to monochromatic radiant exitance $M_{b\lambda}\,(\lambda,T)$ of a black body at the same wavelength, the same temperature and observed at the same angle:

$$\varepsilon_\lambda = \frac{M_\lambda(\lambda, T)}{M_{b\lambda}(\lambda, T)}. \tag{2.24}$$

In terms of surface radiation properties, physical bodies can be divided in the following way (Kreith 2000, Minkina 2004):

- $\varepsilon_b(\alpha) = 1$, $\varepsilon_b(\lambda,T) = 1$, black bodies; α, angle of observation;
- $0 < \varepsilon(\lambda,T) < 1$, non-black bodies;
- $\varepsilon(\alpha) = \text{const}$, $\varepsilon(\alpha) < 1$, dissipative bodies.

Non-black bodies are divided into:

- $0 < \varepsilon(\lambda,T) < 1$, $\varepsilon(\lambda,T) = \text{const}$, $\varepsilon(\alpha) = \text{var}$, gray bodies;
- $0 < \varepsilon(\lambda,T) < 1$, $\varepsilon(\lambda,T) = \text{var}$, $\varepsilon(\alpha) = \text{var}$, non-gray bodies (i.e. selectively emitting bodies).

A dissipative body is a body whose emissivity is independent of angle of observation α. Its surface satisfies the conditions of Lambert's law (so it is called a Lambertian surface). Similarly, we can define a reflective body as a body whose reflectance R is independent of angle of observation α (Touloukian and DeWitt 1970, Touloukian and DeWitt 1972, Touloukian *et al.* 1972).

To investigate how the observation angle affects the radiation properties of a body, we need to introduce some basic laws and definitions associated with the physical aspects of optics (ASTM E 1316).

Lambert's law (cosine law of optics) determines the intensity of radiation emitted by a surface element of a black body versus the distribution angle α (Michalski *et al.* 1991):

$$I_{b\alpha} = I_{b\perp} \cos\alpha, \text{W sr}^{-1}, \tag{2.25}$$

where $I_{b\perp}$ is the radiant intensity emitted in a direction normal to the surface and $I_{b\alpha}$ is the radiant intensity emitted at angle α to the normal to the surface. This equation states that the radiant emissivity from a Lambertian surface is directly proportional to the cosine of angle α

between the observer's line of sight and the normal to the surface. The radiant intensity of a black body surface in the surface normal direction $I_{b\perp}$ is π times smaller than the total radiant intensity I_b emitted from this surface:

$$I_{\mathrm{b}\perp} = I_{\mathrm{b}}/\pi, \mathrm{W\ sr}^{-1}. \tag{2.26}$$

Relationships (2.25) and (2.26) are true also for dissipative bodies. For non-black bodies, formula (2.25) is satisfied only approximately, especially in the case of polished metals and for $\alpha > 50°$. The deviations are due to the dependence of a (real) non-black body emissivity ε_α on angle α.

Luminous intensity I_v is the light flux in a given direction per unit solid angle. Lambert's law holds also with respect to luminous intensity, that is:

$$I_{v\alpha} = I_{v\perp} \cos\alpha, \mathrm{cd}, \tag{2.27}$$

where $I_{v\perp}$ is the luminous intensity in a direction normal to the surface.

Luminance or **brightness L_v** is the surface density of luminous intensity in a given direction:

$$L_v = \frac{dI_v}{dF \cdot \cos\alpha}, \mathrm{cd\ m}^{-2}, \tag{2.28}$$

where dF is the elementary radiant area and α the angle between the observer's line of sight and the normal to the surface (angle of observation).

Luminance describes the subjective impression of a surface brightness. Taking into account (2.27), formula (2.28) takes the form:

$$L_v = \frac{\Delta I_{v\perp} \cdot \cos\alpha}{\Delta F \cdot \cos\alpha} = \frac{\Delta I_{v\perp}}{\Delta F}. \tag{2.29}$$

From (2.29) we can see that the luminance of a black body surface is independent of observation angle α and equal to the luminance in the normal direction. In the case of non-black bodies, usually encountered in reality, luminance is nearly constant for angles α in the interval from zero to $\pi/4$. Apart from the observation angle, the emissivity of a surface depends also on observation time. This is due to variations of the emissivity over time (Madura *et al.* 2001). It turns out that ultrafast thermal phenomena are accompanied by significant changes of emissivity. This effect may result in deterioration of the accuracy of thermography methods applied to ultrafast thermal processes (e.g. active dynamic thermography). Taking into consideration the research so far, we can state that the emissivity of a given body surface is a function of angle of observation α, wavelength λ, body temperature T and time τ:

$$\varepsilon = f(\alpha, \lambda, T, \tau). \tag{2.30}$$

In the case of semitransparent bodies, the emissivity coefficient can be expressed as (Siegel 1992, Linhart and Linhart 2002, Orzechowski 2002):

$$\varepsilon = \frac{(1-R)(1-TT)}{1-RTT} \tag{2.31}$$

To make possible comparisons of material properties independently of the state of its surface, the so-called specific emissivity is sometimes used. It is denoted as ε', total specific emissivity as ε'_{λ}, and monochromatic specific emissivity or $\varepsilon'_{\lambda1\text{-}\lambda2}$ as band-specific emissivity. All these specific emissivities are evaluated in a direction normal to a flat, polished and non-transparent surface. They are also known and published as normal emissivities, that is emissivities evaluated along the normal to the surface $\alpha = 0$ (see the table of normal emissivities of various materials in Appendix B).

For reliable estimation of emissivity many other factors should be taken into account, such as the condition of a considered object surface or its homogeneity. These are very difficult to describe mathematically, so values of emissivity for particular bodies are usually determined with low accuracy. This turned out to be a problem in infrared thermography because setting an accurate value for the considered object emissivity in the mathematical model of an infrared camera measurement path is of great importance for the correct evaluation of its temperature. Therefore, evaluation of the observed surface emissivity ε as accurately as possible should be an important stage of each measurement in infrared thermography.

2.3.2 Methods for Evaluating Emissivity

There are various methods for the evaluation of emissivity. For example, Orlove (1982) proposes the following scheme:

- Stick on the object's surface a piece of material (sticker) of high and accurately known emissivity (e.g. $\varepsilon = 0.95$) and of good thermal conductivity, or paint part of it with a special paint of known and high emissivity.
- Heat up the object to a temperature at least 50 °C higher than the ambient temperature.
- Set up the camera spot point (SP) on the part of the object with the sticker (or previously painted).
- Set on the camera the known emissivity of the sticker (or the paint) and measured earlier values of atmospheric temperature, the ambient temperature, the camera-to-object distance and atmospheric humidity.
- Read the spot point temperature of the area of known emissivity.
- Move the spot point outside the area of known emissivity.
- Change the parameter of the object emissivity in the camera and read the spot point temperature until it is the same as for the 'clean' area of known emissivity.

A variant of this method consists of determining the object's temperature by using a contact method. Then, the emissivity parameter in the camera should be tuned until the same reading of temperature is obtained. The value of the parameter set last represents the object's emissivity. In another method, a hole of depth at least six times larger than its diameter is drilled in the surface. Such a hole can be treated as a black body of emissivity $\varepsilon_{ob} \approx 1$. This method is an approximate method, since the hole distorts the temperature field of the object's surface.

The emissivity coefficient ε_{α} of a high-temperature object (it depends on angle of observation α), for an arbitrary point on a curved cylinder or the emissivity of an arbitrary flat surface, can also be evaluated in the way described below.

According to the formula derived in 1883 by C. Christiansen, the heat flux exchanged between surfaces 1 and 2, where surface 1 has area F_1 much smaller than area F_2 of surface 2, is given as:

$$\Phi_{2-1} = F_1 \cdot \varepsilon_1 \cdot C_o \left[\left(\frac{T_2}{100} \right)^4 - \left(\frac{T_1}{100} \right)^4 \right], \mathrm{W}. \tag{2.32}$$

Hence, the heat flux of a black body surface at temperature T_{bb}, arriving at the camera detector at temperature T_d and area F_d, is:

$$\Phi_{1-2} = \sigma_o \cdot F_d (T_{bb}^4 - T_d^4), \mathrm{W}. \tag{2.33}$$

Formula (2.33) is only an approximation, because it does not take into account the geometry of the camera lenses and parameters of the atmospheric transmission.

The heat flux emitted by a non-black body of angle-dependent emissivity ε_α under the same conditions is:

$$\Phi_{1-2} = \sigma_o \cdot F_d \cdot \varepsilon_\alpha \cdot (T_{bb}^4 - T_d^4), \mathrm{W}. \tag{2.34}$$

When we enter emissivity $\varepsilon_\alpha = 1$ into the camera, it will show some temperature T_s, different (lower) than T_{bb}. However, the same heat flux:

$$\Phi_{1-2} = \sigma_o \cdot F_d \cdot (T_s^4 - T_d^4), \mathrm{W} \tag{2.35}$$

still arrives at the detector. Comparing the right-hand sides of (2.34) and (2.35), we obtain the following relationship for ε_α as a function of temperatures T_d, T_{bb} and T_s:

$$\varepsilon_\alpha = \frac{T_s^4 - T_d^4}{T_{bb}^4 - T_d^4}. \tag{2.36}$$

When T_s (T_{bb}) is much higher than the camera detector temperature T_d (for cameras with cooled detectors $T_d \approx 70$–$200\,\mathrm{K}$), formula (2.33) can be approximated by:

$$\varepsilon_\alpha \approx \left(\frac{T_s}{T_{bb}} \right)^4. \tag{2.37}$$

The emissivity determined in this way is an average emissivity within the detector's sensitivity band.

For other materials one of the following approximate relationships can be used:

- The emissivity of a perfectly smooth metal surface as a function of wavelength λ (the relationship holds for $\lambda > 2\,\mu\mathrm{m}$; Sala 1993):

$$\varepsilon = k \sqrt{\frac{\rho}{\lambda}}, \tag{2.38}$$

  where $k = 0.365\,\Omega^{-1/2}$ is a constant coefficient and ρ the resistivity (Ω m).
- The emissivity of a real metal surface as a function of wavelength λ (Sala 1993):

$$\varepsilon = \frac{1}{b_1 \sqrt{\lambda} + b_2}, \tag{2.39}$$

  where b_1 $\mu\mathrm{m}^{-1/2}$ and b_2 are constant coefficients.

- The monochromatic emissivity ε_λ of non-conducting materials versus refractivity n_λ (Michalski *et al.* 1991, Michalski *et al.* 1998):

$$\varepsilon_\lambda = \frac{4n_\lambda}{(n_\lambda + 1)^2}, \tag{2.40}$$

where $n_\lambda = 1.5$–4 for non-organic compounds and 2.0–3.0 for metal oxides.

Other methods for the evaluation of emissivity can be found in Marshall (1981), Madding (1999) and Madding (2000).

2.3.3 Examples of Evaluating Emissivity and Its Effect on Temperature Measurement

Example 2.3 *Experimental evaluation of the surface emissivity of a central heating (CH) radiator under stationary conditions of heat exchange*

In work by Dudzik (2007), Dudzik and Minkina (2008a) and Dudzik (2008), the following procedure was applied to evaluate the surface emissivity of a panel CH radiator:

- definition of coordinates of measurement points on the heater surface;
- measurement of temperature at the defined points using a contact method (Minkina 1992, Minkina 1999, Minkina and Gryś 2002c) accompanied by simultaneous recording of temperature in the close neighborhood of the measurement points using the thermography method;
- application of the least squares method to evaluate the heater surface emissivity.

The photograph in Figure 2.4 shows the one-row panel heater with a single convective part, of normal power 980.53 W, produced by DeLonghi. The thermogram of the heater

Figure 2.4 Photograph of the heater with installed temperature sensors LM35A

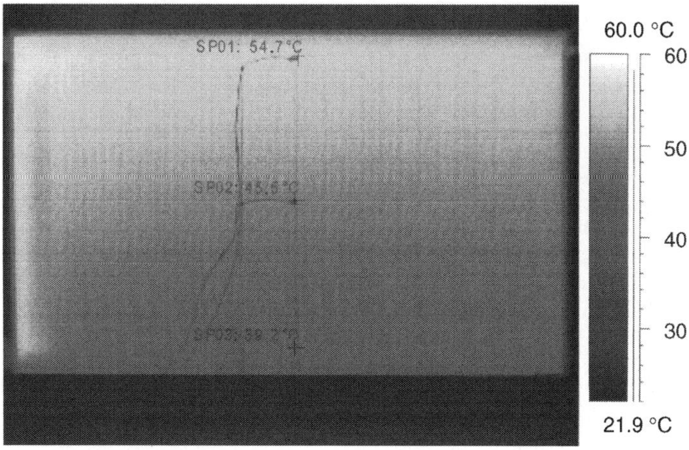

Figure 2.5 Thermogram of the heater with marked temperature measurement points. See Color Plate 1 for the color version

presented in Figure 2.5 was recorded on a laboratory stand with an open chamber. The chamber walls were covered with an emulsion of high emissivity ($\varepsilon \approx 1$). Next, 60 thermograms recorded during the test were averaged. The heater was supplied with water at a controlled inflow temperature. All the thermograms were recorded under stationary thermal conditions for the following parameters of heating medium (water):

- inflow temperature: (58.67 ± 0.01) °C;
- outflow temperature: (48.16 ± 0.01) °C;
- volumetric flow: $(47 \pm 3)\ \mathrm{l\,h^{-1}}$.

The values of temperature recorded in the neighborhood of the measurement points were read from the averaged thermograms. The LM35A temperature sensors were mounted at points with the following coordinates: $x_1 = 165$ pixels, $y_1 = 29$ pixels, $x_2 = 165$ pixels, $y_2 = 106$ pixels, $x_3 = 165$ pixels, $y_3 = 185$ pixels, where the numbers denote coordinates of the 320×240 pixel camera array detector. The measurement conditions are given in Table 2.1.

Below we present example values of compensated signals, calculated from the thermogram in close neighborhood to the measurement points using formula (3.13):

$$s_{ob_1} = 33\,231, \text{for } h_1 = (0.55 \pm 0.002)\ \text{m, at } T_{ob_1} = (53.9 \pm 0.5)\ ^\circ\text{C},$$

$$s_{ob_2} = 32\,963, \text{for } h_2 = (0.30 \pm 0.002)\ \text{m, at } T_{ob_2} = (45.1 \pm 0.5)\ ^\circ\text{C},$$

Table 2.1 Conditions of the infrared thermography measurement for evaluation of emissivity of the CH heater surface

Ambient temperature, °C	Temperature of atmosphere, °C	Emissivity	Relative humidity, %	Camera-to-object distance, m
20 ± 0.2	20 ± 0.2	—	0.5	0.252 ± 0.001

$$s_{ob_3} = 32\,796, \text{for } h_3 = (0.05 \pm 0.002)\ \text{m, at } T_{ob_3} = (40.1 \pm 0.5)\ °\text{C}.$$

Provided that the inputs T_{atm}, T_o, ω and d are constant, formula (3.17) of the temperature measurement model can be written as:

$$\hat{T}_{ob} = f(s_{ob}, \varepsilon_{ob}), °\text{C}, \tag{2.41}$$

where: $\hat{T}_{ob}$ is the object temperature calculated from the model equation; s_{ob} is the value of the detector's compensated signal – see (3.13); and ε_{ob} is the object emissivity.

Denoting the temperature of the object measured using a contact method by T_{ob}, the value of emissivity was determined so that the measured and calculated temperatures were very similar (Minkina and Chudzik 2003):

$$\begin{aligned} D_1 &= \hat{T}_{ob_1} - T_{ob_1} \approx 0 \\ D_2 &= \hat{T}_{ob_3} - T_{ob_2} \approx 0 \\ D_3 &= \hat{T}_{ob_3} - T_{ob_3} \approx 0 \end{aligned} \tag{2.42}$$

requiring that the sum:

$$g(\varepsilon_{ob}) = \sum_{i=1}^{3} [\hat{T}_{ob_i}(\varepsilon_{ob}) - T_{ob_i}]^2, \tag{2.43}$$

reaches a minimum. Minimization of (2.43) was carried out using the Levenberg–Marquardt optimization algorithm (Nelles 2001).

The values of the measured and calculated temperatures using the least squares method, at the three considered points, are presented in Figure 2.6. The corresponding differences between the measured and calculated temperatures are shown in Figure 2.7.

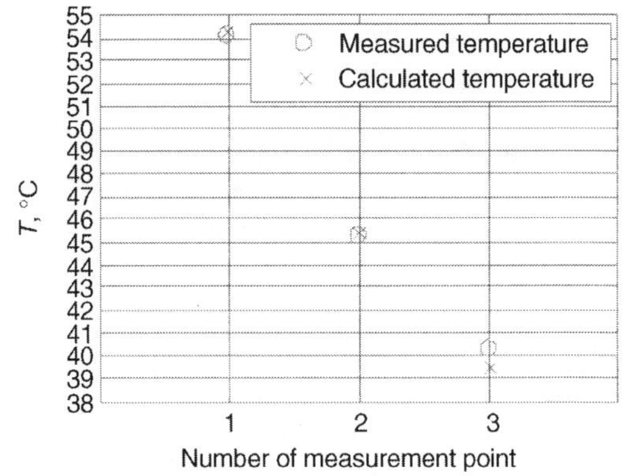

Figure 2.6 The values of temperatures measured and calculated using the least squares method, at the three measurement points of the heater

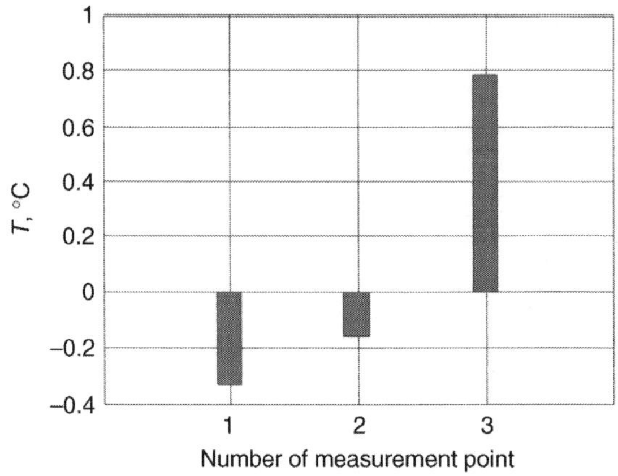

Figure 2.7 Differences between the measured and calculated temperatures at the three considered points

The limiting error for evaluating the emissivity was calculated in the following way:

- The limiting error of the contact measurement was assumed equal to ±0.5 °C (Bernhard 2003).
- The negative limiting error of the emissivity measurement was determined by increasing the values of limiting temperatures at the three points by 0.5 °C (i.e. by the positive limiting error of the contact measurement) and evaluating a new (underrated) value of emissivity.
- The positive limiting error of the emissivity measurement was determined by decreasing the values of limiting temperatures at the three points by 0.5 °C (i.e. by the negative limiting error of the contact measurement) and evaluating a new (overrated) value of emissivity.

As a result of this procedure, we obtained the emissivity of the heater surface as equal to $\varepsilon_{ob} = 0.97 \pm 0.02$. The value of the limiting error (0.02) is about 2.1% of the evaluated value of the emissivity. It should be emphasized that the worst possible measurement conditions were assumed for estimation of the error, that is:

- maximum limiting error of the LM35A sensors: according to the catalog, this is equal to 0.5 °C (typical error is 0.3 °C) (Lieneweg 1976, Minkina and Gryś 2005);
- worst distribution of the limiting errors of the sensors: the errors for all sensors were taken with the same sign.

In reality, it is unlikely that the temperature measurement at every point is burdened with the maximum error of the same sign (Taylor 1997, Quinn 1983). Therefore, the error of the emissivity evaluated according to this procedure is probably overestimated.

Example 2.4 *How the change in the measured temperature affected the change in the emissivity*

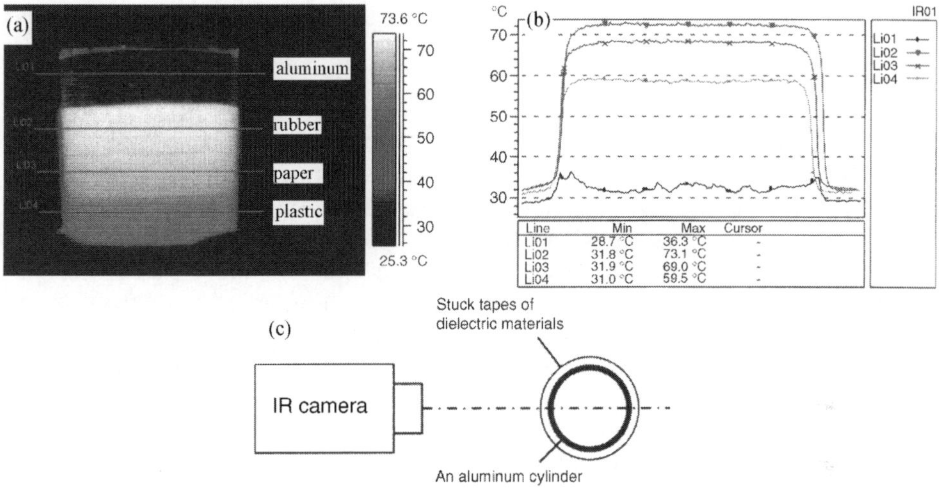

Figure 2.8 Infrared thermography temperature measurement of an aluminum cylinder (cross-section Li01) with stuck tapes of dielectric materials: rubber (cross-section Li02), paper (cross-section Li03) and plastic (cross-section Li04) under stationary conditions. Infrared camera shows an apparently different temperature for each material: (a) thermogram; (b) temperature profiles; and (c) top view of the experimental setup (Minkina2004). See Color Plate 2 for the color version. Reproduced by permission of Częstochowa University of Technology

The dependence of the temperature measurement outcome on emissivity ε and angle of observation α is clearly visible in the image and graph in Figure 2.8. The thermogram shows a cylinder made of aluminum sheet ($\varepsilon_{ob} = 0.09$), cross-section Li01 with stuck tapes of dielectric materials; rubber ($\varepsilon_{ob} = 0.95$), cross-section Li02; paper ($\varepsilon_{ob} = 0.92$), cross-section Li03; and plastic ($\varepsilon_{ob} = 0.87$), cross-section Li04. The cylinder temperature was fixed and the same over the whole surface. It was filled with water of temperature about 80 °C. In the experiment the following measurement parameters were introduced into the infrared camera microcontroller: $\varepsilon_{ob} = 1$; $T_o = T_{atm} = 24\,°\text{C}$ (297.15 K); $\omega = 0.5$; $d = 0.6$ m. However, the camera sees an apparently different temperature for each material. The highest temperature is shown for the rubber tape, because rubber has the greatest emissivity of the four materials. On the other hand, the aluminum surface is interpreted to be the coolest because its emissivity is lowest.

Since measurements of emissivity are frequently burdened with significant errors, calibration of infrared cameras is of great importance. This issue is dealt with further in Chapter 4. Next, we present basic information on infrared cameras with particular reference to measurement cameras.

2.4 Measurement Infrared Cameras

The basic component of an infrared system is an infrared camera. Because the atmosphere has two bands of good transmission in the infrared range (i.e. short-wave band between 2 and 5 μm and long-wave band between 8 and 14 μm), most detectors and infrared (IR) cameras are divided in a natural way into short-wave (SW) and long-wave (LW) devices. However, there are detectors which work in the near IR (0.78–1.5 μm), for example quantum and photoemissive detectors, and detectors which work in the far IR (20–1000 μm), for example thermal detectors (Maldague 2001). Another classification follows from the detector type: there are cameras with cooled detectors, containing a refrigerator (cooling) unit, and non-cooled detectors, operating in the ambient temperature. Until 1997 all the IR cameras produced were equipped with detectors cooled to temperatures from −70 (seldom) to −200 °C (most frequently). Manufacturers offer measurement IR cameras (calibrated by the manufacturer), used for temperature measurement, and imaging IR cameras that show only a color map of the approximate temperature field. Imaging cameras are cheaper, so they are used more often, for example by border officials or police for night surveillance.

Detectors in IR cameras are divided into: point (single) detectors, linear and array detectors (FPA, Focal Plane Array) built up as matrices consisting of, for example, 640 × 480 individual detectors (pixels).

Cameras with a single detector or a ruler of detectors are sometimes called point (single) detector scanners or linear scanners, respectively. In such cameras the image of the temperature field is created by an optomechanical scanning system built of rotating or oscillating mirrors, or scanning prisms. The scanning frequency is usually equal to 25 Hz (50 Hz) for the PAL system in Europe, or 30 Hz (60 Hz) for the NTSC system in the USA.

In a single detector camera the image of the observed area is built point by point at consecutive time instants. Radiation arriving at the detector is converted into electrical signals proportional to the radiant exitance of individual points of the image. The signals are amplified and transmitted synchronously with a scanning motion to a display (formerly a scope), where the temperature field image (thermogram) is created. This principle of operation was in use for 20 years after the first camera came out. Systems had one detector, whose characteristic determined the type of scanner and its thermal and spatial resolution, that is its capability of distinguishing temperature at two adjacent points and the number of pixels in a thermogram respectively. Single detector scanning cameras have unique metrological properties. All points of a thermogram have identical parameters because the temperature at every point is measured by the same detector. This is particularly significant in detecting the temperature difference at two points of a homogeneous object. Such a camera can better perform self-calibration before each measurement, compensating, for example, changes in the detector's sensitivity or changes in the amplification of electronic circuits (Machin and Chu 2000, Machin *et al.* 2008). It is also easier to design and make lenses that do not introduce optical or energy distortions (De Mey 1989, De Mey and Wiecek 1998).

Construction of rulers of detectors and linear scanner cameras (Figure 2.9c) was the next step in the development of thermal imaging systems. Such systems have one scanning unit, either vertical or horizontal depending on the detector's ruler mounting.

Since 1993 cameras have been equipped more and more frequently with FPA detectors. A typical 640 × 480 array (matrix) is built up of 307 200 individual detectors (pixels). Each pixel is read 25 (50) (PAL system – Europe) or 30 (60) (NTSC system – USA) times per second by the ROIC readout system. The array readout frequency is published in catalogs as 'image frequency'. Arrays are available containing different numbers of detectors. In cameras with array detectors there are no mechanical scanning parts: the matrix 'looks' at an object through the camera optics (Figure 2.10). The development of fast array detectors enabled the construction of cameras capable of recording ultrafast thermal processes and the rise of a new branch of IR thermography measurements called ultrafast thermography.

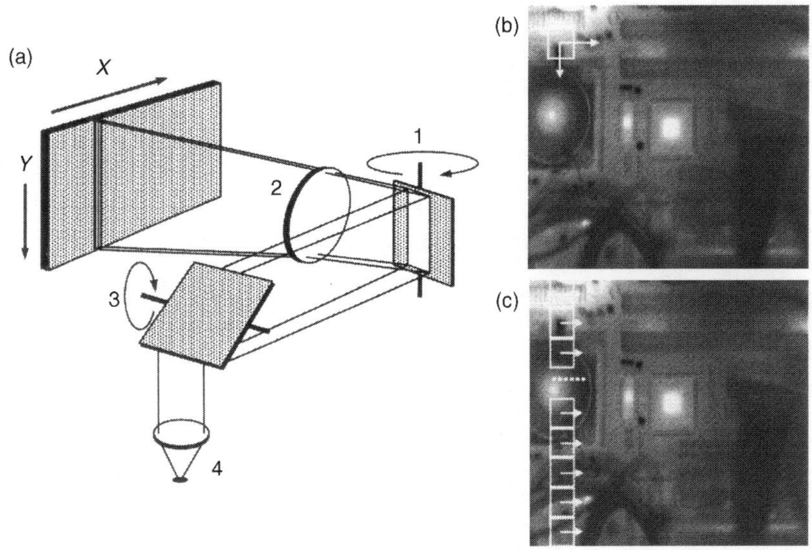

Figure 2.9 Creation of thermogram in (a) point (single) detector and (b) linear scanner: 1, horizontal deflection mirror; 2, optics; 3, vertical deflection mirror; 4, point (single) detector (b), linear detector (c)

Currently there are infrared systems capable of recording several hundred thermograms per second.

The next step in the development of IR cameras was the introduction in 1997 of the first camera with a microbolometric array of non-cooled detectors. Somewhat later, non-cooled arrays built of pyroelectric detectors were constructed. The elimination of mechanical scanning and cooling improved the operating parameters of IR cameras, which became lighter, more reliable and able to work much faster. Cooling the detector to cryogenic temperature took over 10 minutes, whereas stabilizing the operating temperature in a camera without a cooler does not exceed 1 minute.

Both measurement and imaging IR cameras are characterized by many parameters describing their imaging and measurement properties (ASTM E 12213, ASTM E 1311). Since this

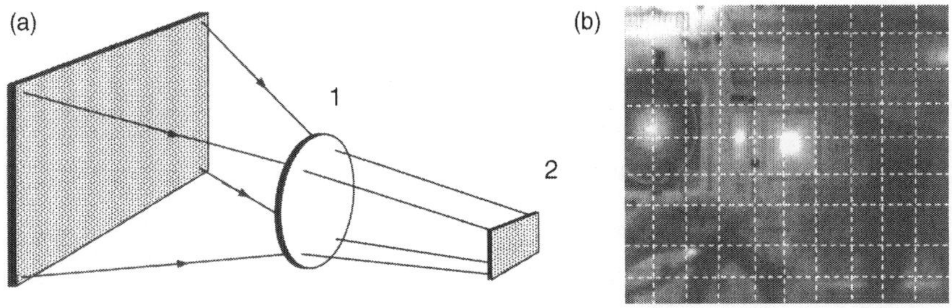

Figure 2.10 Recording of thermogram in a focal plane array (FPA) camera: 1, camera optics; 2, array of detectors

book concerns errors and uncertainties of measurements in IR thermography, we concentrate on the metrological properties of measurement cameras. Imaging properties of IR cameras are described in detail in, for example, Minkina (2004), Chrzanowski (2000) and Nowakowski (2001). Below we list and describe the most important parameters of contemporary measurement IR cameras.

2.4.1 Noise Equivalent Temperature Difference (NETD)

This is the difference between the temperature of the observed object and the ambient temperature that generates a signal level equal to the noise level. It is also called the temperature resolution. NETD is defined as the ratio of the RMS noise voltage U_n to the voltage increment ΔU_s generated by the difference in temperature between the measurement area of a technical black body (or test body) T_{ob} and background temperature T_o, divided by this difference:

$$\mathrm{NETD} = \frac{U_n}{\dfrac{\Delta U_s}{T_{ob} - T_o}} = \frac{T_{ob} - T_o}{\dfrac{\Delta U_s}{U_n}}, \mathrm{K}. \tag{2.44}$$

The temperature of a technical black body measurement area is usually equal to 30 °C, with a background temperature of 22 °C, and the difference $T_{ob} - T_o$ should be within 5–10 K (Figure 2.11).

There is also another, slightly different definition of the NETD parameter: it is defined as the difference of temperatures T_{ob} and T_o observed by the detector, which results in a change in the output signal equal to the noise of the detector. NETD is determined from observation of the area of a technical black body, whose temperature T_{ob} is close to the background temperature T_o (Figure 2.11a). An example of a signal coming from a detector along line N is presented in Figure 2.11b. The value of NETD is determined when signal U_s is equal to noise level U_n.

In both cases NETD is defined as the minimum increment of the temperature difference, or as the minimum difference of temperatures T_{ob} and T_o that can be distinguished by a point (single) detector (or a linear or array detector) for a given amplifier bandwidth. According to the theory presented in Bielecki and Rogalski (2001), narrowing of the amplifier bandwidth results in a decrease of the noise voltage (i.e. a decrease of NETD), but on the other hand it deteriorates the spatial resolution (e.g. for a constant scanning speed). Circular or rectangular test fields of stabilized temperature T_{ob} can be used in measurements instead of a technical black body measurement field. The presented definitions of NETD do not take into consideration the size

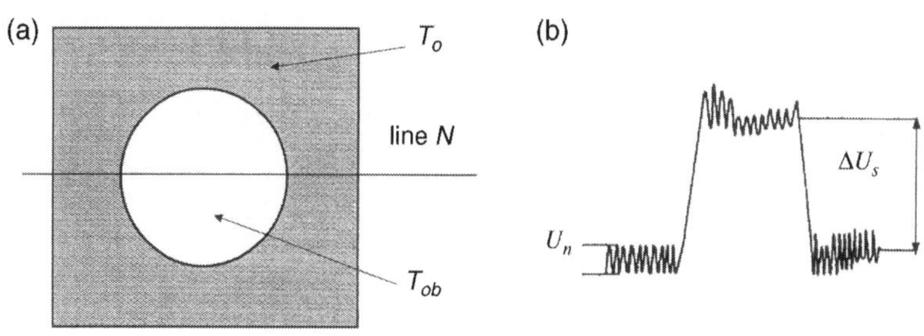

Figure 2.11 Interpretation of determination of the NETD parameter (Nowakowski2001)

of an object, physiology of human perception or properties of the display system. To appreciate better how the NETD parameter is evaluated, we present below an example based on the work of Minkina (2004).

Example 2.5 *Comparison of Noise Properties of two Infrared Systems*

Let us determine which of two systems has the smaller noise voltage U_n. We assume identical increments of the detector signal ΔU_s for both cameras. The background temperature T_o is 22 °C. Let us also assume that the values of the NETD coefficients of the cameras given in the technical specifications are as follows:

- $\mathrm{NETD}_1 = 0.1$ K for $T_{ob1} = 30\,^\circ\mathrm{C}$;
- $\mathrm{NETD}_2 = 0.2$ K for $T_{ob2} = 50\,^\circ\mathrm{C}$.

According to formula (2.44) we obtain:

$$U_n = \frac{\mathrm{NETD}}{T_{ob} - T_o}\Delta U_s. \tag{2.45}$$

Hence:

$$\begin{aligned} U_{n1} &= \frac{\mathrm{NETD}_1}{T_{ob1} - T_o}\Delta U_s = \frac{0.1 \cdot \Delta U_s}{8} = 0.0125 \cdot \Delta U_s, \\ U_{n2} &= \frac{\mathrm{NETD}_2}{T_{ob2} - T_o}\Delta U_s = \frac{0.2 \cdot \Delta U_s}{28} = 0.007 \cdot \Delta U_s. \end{aligned} \tag{2.46}$$

We conclude that camera 2 has a smaller value of the noise voltage, even though its NETD coefficient is larger ($\mathrm{NETD}_2 > \mathrm{NETD}_1$).

A simplified method of NETD measurement can be found at www.vigo.com.pl. Measurements are carried out for the first measurement range of a camera, and temperature T_{ob} of the observed object is set so that it is in the middle of this range. The object temperature is assumed to be known, constant and homogeneous. On the recorded thermogram we choose a line of the temperature distribution profile that crosses the geometrical center of the object, as in Figure 2.11a, and find on this line an interval in which the average temperature is constant. If we denote the highest and the lowest temperature indicated in this interval by $T_{\max}$ and $T_{\min}$ respectively, NETD can be determined as:

$$\mathrm{NETD} = \frac{T_{\max} - T_{\min}}{2}, \mathrm{K}. \tag{2.47}$$

The measurement should be repeated for several lines. It is also possible to select on the thermogram areas whose average temperature distribution is constant. NETD is then the mean of these measurements.

To determine how NETD depends on temperature T_{ob}, it should be set once near the lower and then near the upper limit of the camera's measurement range. Examples of the relationship of NETD to T_{ob} for an FLIR SC 3000 camera with a QWIP detector are shown in Figure 2.12. Analogous graphs for an SW and LW camera are compared in Figures 2.13a and b. In

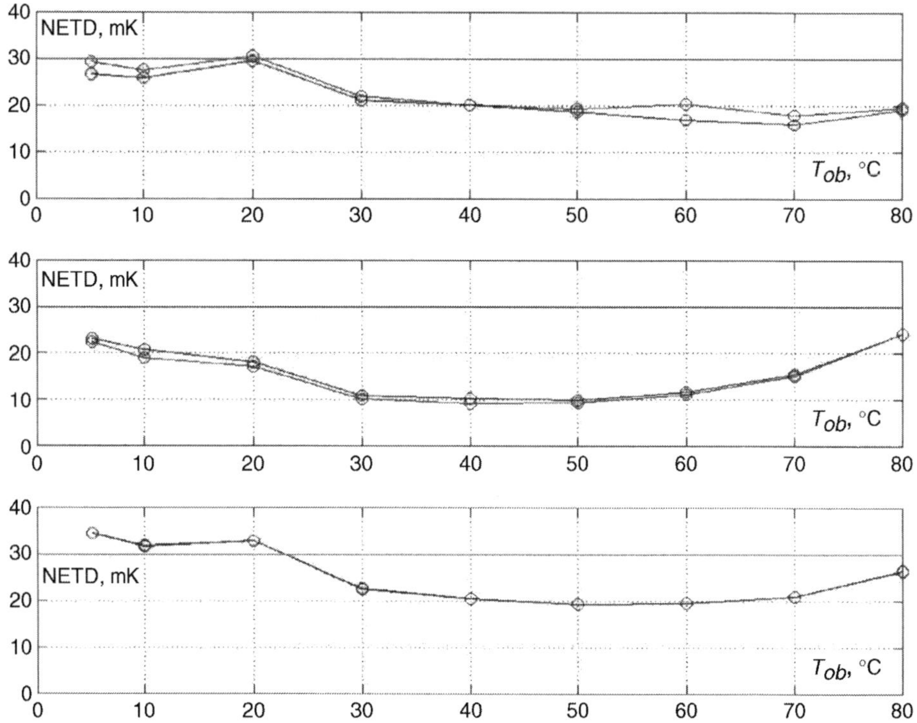

Figure 2.12 Temperature resolution NETD versus T_{ob} for the FLIR SC 3000 camera with a QWIP detector, obtained from a series of measurements (QWIP Seminar2000)

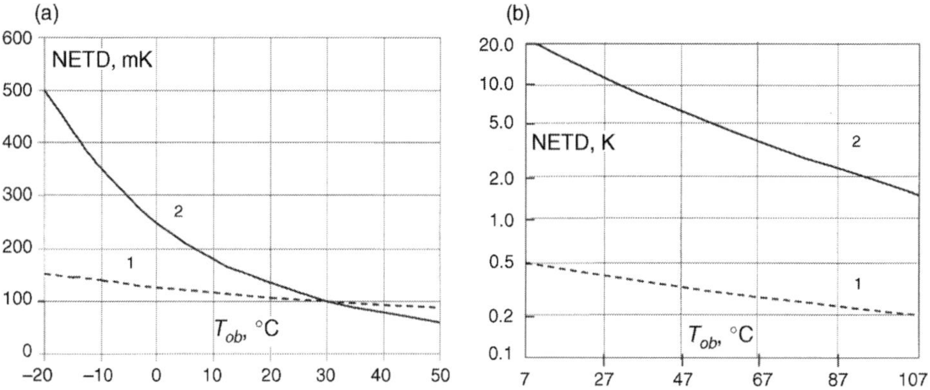

Figure 2.13 Typical graphs of temperature resolution NETD versus T_{ob} for: (a) LW camera (1) and SW camera (2) (IR-Book2000); (b) Inframetrics 760 BB camera with filters, operating in LW range 8–12 μm (1) and SW range 3–5 μm (2) (Chrzanowski2000)

Figure 2.13a, it is assumed for both types of cameras that NETD = 100 mK for T_{ob} = 30 °C. Notice the significant sensitivity of NETD to T_{ob}, especially for SW cameras. A greater value of NETD indicates a lower sensitivity of the camera. Therefore, in the technical data of IR cameras, the NETD parameter is called 'thermal sensitivity' or 'temperature resolution'. The catalog value of the temperature resolution should be accompanied by the value of T_{ob} at which NETD was evaluated. The temperature resolution for different types of cameras has the following typical values:

- 10–30 mK – for cameras with QWIP detectors, designed for research and development applications;
- 50–100 mK – for measurement cameras;
- >200 mK – for imaging cameras.

Procedures for evaluating the temperature resolution are not yet standardized. Therefore, the values of NETD published in technical data can be evaluated in different ways and, consequently, are incomparable. NETD is one of the most popular parameters for assessing the metrological properties of an IR camera. As a result, a worse camera might be characterized by a better (i.e. lower) value of NETD for marketing purposes. This parameter is freely applied by camera manufacturers for assessment of the cameras' metrological properties.

It should be noted that, if the spectral characteristic of the detector's normalized detectivity is known, the NETD parameter can be evaluated theoretically. However, this is a broad issue that goes beyond the scope of this monograph (Madura *et al.* 2004).

2.4.2 Field of View (FOV)

This determines the area that can be observed from a given distance *d* using the optics installed on the camera. This parameter determines the spatial (geometrical) resolution of a measurement IR camera. FOV is defined in meters and determines resolution in both the horizontal (*H*) and vertical (*V*) directions. Typical values of FOV as a function of distance *d* for 24° × 18° optics are given in Table 2.2. FOV for these optics is calculated as:

$$H = d \cdot \sin 24°, \text{m}; \quad V = d \cdot \sin 18°, \text{m} \tag{2.48}$$

2.4.3 Instantaneous Field of View (IFOV)

This determines the FOV of a single detector (pixel) in an array. From a practical point of view, it should be called the 'minimal field of view'. It is the second parameter determining the spatial (geometrical) resolution of a measurement IR camera. In technical data it is referred to as the

Table 2.2 Field of view (FOV) versus distance d for 24° × 18° optics (Minkina2001)

d, m	0.50	1.0	2.0	5.0	10	30	100
H, m	0.20	0.41	0.81	2.0	4.1	12	41
V, m	0.15	0.31	0.62	1.5	3.1	9.3	31

'spatial resolution'. For example, FOV $H \times V$ at distance $d = 1$ m for a camera with $24° \times 18°$ optics is 0.41×0.31 m (see Table 2.2). If the camera has a matrix of 320×240 detectors, the FOV of a single detector $H_{min} \times V_{min}$ is:

$$\frac{0.41}{320} = \frac{0.31}{240} = 1.3\,\text{mm} \cdot 1.3\,\text{mm}.$$

This means that at distance of 1 m such a camera can recognize details like local overheats 1.3 mm by 1.3 mm in size. Obviously, it is capable of detecting an overheat on a smaller area, but in such a case the measured temperature will be underrated. The IFOV parameter is proportional to the distance, so for $d = 10$ m in the above example it will be 13 mm by 13 mm.

Another method of evaluating IFOV is based on determining the optical angle of flare (in radians) α_{rad} for a single detector:

$$\alpha_{radH} = \frac{24\pi}{180 \cdot 320} = 0.0013\,\text{rad}, \tag{2.49}$$

thus $H_{min} = d{\cdot}\sin 0.0013) = 1.3$ mm,

$$\alpha_{radV} = \frac{18\pi}{180 \cdot 240} = 0.0013\,\text{rad} \tag{2.50}$$

and $V_{min} = d{\cdot}\sin(0.0013) = 1.3$ mm.

In other words, IFOV is the area that a single pixel 'looks at' through the camera optics, and it determines the absolute lower limit of a measured object size. Spatial resolution of a camera depends on applied optics and on the number of detectors (pixels) in the array. The smaller the optical angle of flare and the larger the number of pixels, the better the camera spatial resolution (i.e. the smaller the objects it can observe). However, there are obvious limitations on the size of array detectors in pixels and production of lenses with smaller angles (i.e. smaller FOV).

Spatial resolution of a camera is usually expressed in milliradians (mrad). For example, if technical data specify a spatial resolution of 1.3 mrad for $24° \times 18°$ optics, this parameter is the IFOV (H_{min} and V_{min}) of a single detector (pixel) calculated according to formulas (2.49) and (2.50).

A more detailed analysis of this situation is illustrated in Figures 2.14a and b. It shows two cases of exposure of an array detector to radiation from a small object. In Figure 2.14b the object fully irradiates at least one detector, while in Figure 2.14a the object does not fully irradiate any single detector. In Figures 2.14c and d we can see that the maximum temperature of an anchor clamp measured from a long distance (Figure 2.14c) is lower than that measured from a short distance (Figure 2.14d). This is due to the fact that the object situated closer to the camera fully irradiates one or more detectors in the array. When the object is warmer than the background, its temperature will be underrated if no detector is fully irradiated. Otherwise, for example when the background is warmer, the object temperature will be overrated. The term 'instantaneous' means that fulfillment of the IFOV parameter requirements is sufficient for correct irradiation of a detector theoretically only at some instant and for ideal electronics and optics. In the spatial domain, 'instantaneous' denotes the point response of a detector to the radiant exitance of an object. The symbols in Figures 2.14c and d have the following meaning:

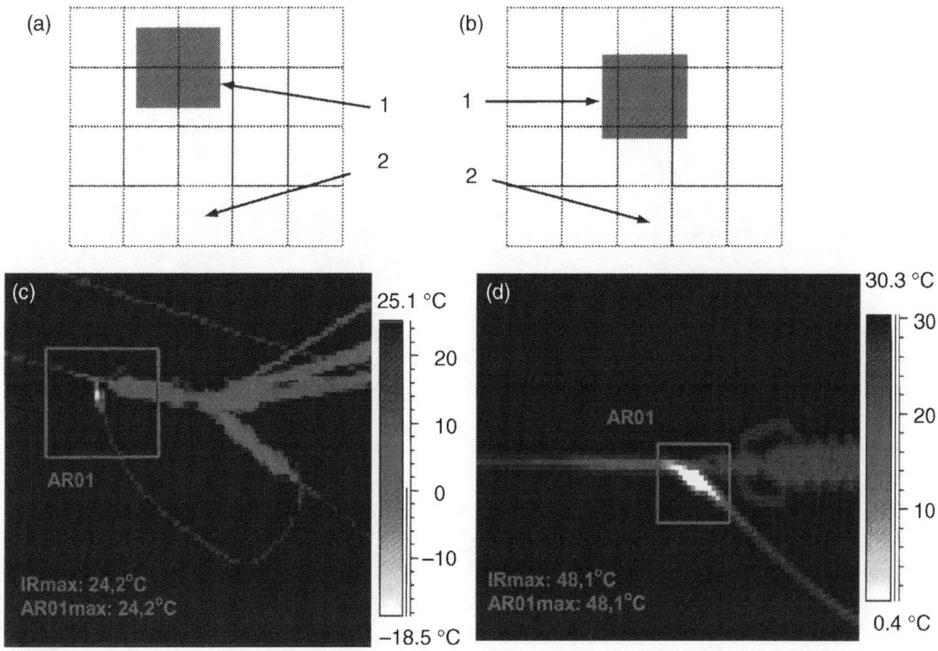

Figure 2.14 Image of a small object (anchor clamp for bridge connection of high-voltage line anchor support) on an array detector allowing correct temperature measurement: (a) no single detector is fully irradiated; (b) at least one detector is fully irradiated (1, object; 2, array detectors); (c) thermogram of a clamp recorded at long distance – about 40 meters (optical and digital zoom); (d) thermogram of a clamp recorded at short distance – about 7 meters (only digital zoom) (Minkina2001). See Color Plate 3 for the color version

- IRmax – maximum temperature of the whole thermogram area, that is maximum indication selected from indications of all detectors in the array;
- ARmax – maximum temperature of the selected thermogram area (marked with the rectangle), that is maximum indication selected from indications of detectors in the selected area.

In Figures 2.14c and d, IRmax = ARmax, which shows that the locations of the thermograms' maximum temperatures are correct. From Figure 2.14b we can conclude that for correct measurement of point temperature, the object should irradiate at least an area of 2×2 pixels (then at least one detector is fully irradiated). In practice, this may not be enough if the object's shape is not square or rectangular. In addition, each real optic distorts an image. These distortions result, for example, from chromatic and/or spherical aberration and many other imperfections of the optics – Figure 2.15. They can be described in a simple way by a **point spread function (PSF)**. One of the popular models is given as:

$$\mathrm{PSF}(x, y) = \exp\left(-\frac{x^2 + y^2}{2\sigma^2} \right), \tag{2.51}$$

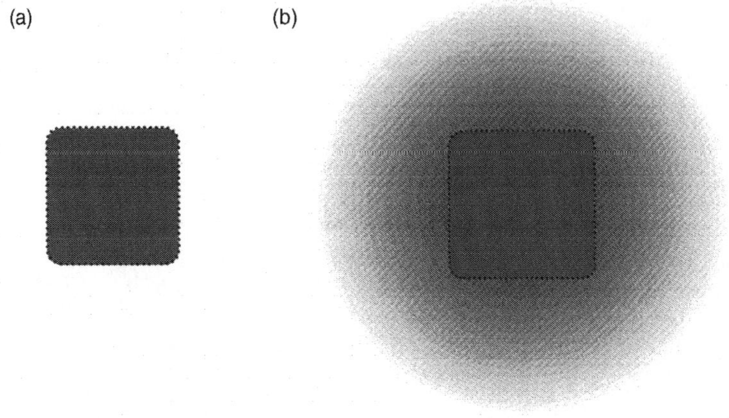

Figure 2.15 Determination of measurement area size: (a) ideal optics – irradiation of area of 2×2 detectors required for correct measurement; (b) real optics, image blurring – irradiation of area of 3×3 or 4×4 (sometimes 5×5) detectors required for correct measurement (Danjoux2001, Minkina2004). See Color Plate 4 for the color version

where σ is a parameter determining the point response of an optical system (spatial resolution) in milliradians. Its typical values are:

- $\sigma = 0.5$ mrad for measurement cameras of better spatial resolution;
- $\sigma = 1.0$ mrad for imaging cameras of lower spatial resolution.

The **slit response function (SRF)** is a parameter that, similar to IFOV, describes the capability of a camera with an array detector to measure the temperature of small objects.

In Figure 2.16 there are three situations that can occur during observation of a black body of temperature T_{ob} through a vertical slit. An area of a black body surface 'seen' by a single detector is denoted by IFOV. Temperature T_{ob} corresponds to signal s_{ob} from the detector. The measurement field is gradually covered by a diaphragm of temperature T_o corresponding to signal s_o from the detector. The value of s_{ob} changes (decreases) with decreasing slit width δ.

Left parts of characteristics from Figure 2.16 are shown in Figure 2.17 (Danjoux 2001) (as a function of optical angle of flare α_{rad}) for a single detector defined by (2.49) and (2.50). The value of α_{rad} is approximately equal to the ratio of slit width δ to camera-to-object distance d. Graph 1 in Figure 2.17a describes an ideal case and graph 2 describes a real case, taking into account that both the optics and the electronics of a camera are not ideal.

A comparison of the real modulation characteristics of a detector's slit response versus the optical angle of flare of a single detector for a measurement camera (curves 1, 2) and an imaging camera (curve 3) is shown in Figure 2.17b. We can see that the value of angle α_{rad} corresponding to 50% modulation is much smaller for a measurement camera, whose slit response function is steeper. Taking into account the remarks we made when discussing the IFOV parameter, we can conclude that the bigger the value of the modulation function, the more accurate the temperature reading. For example, the value equal to 90% means that the

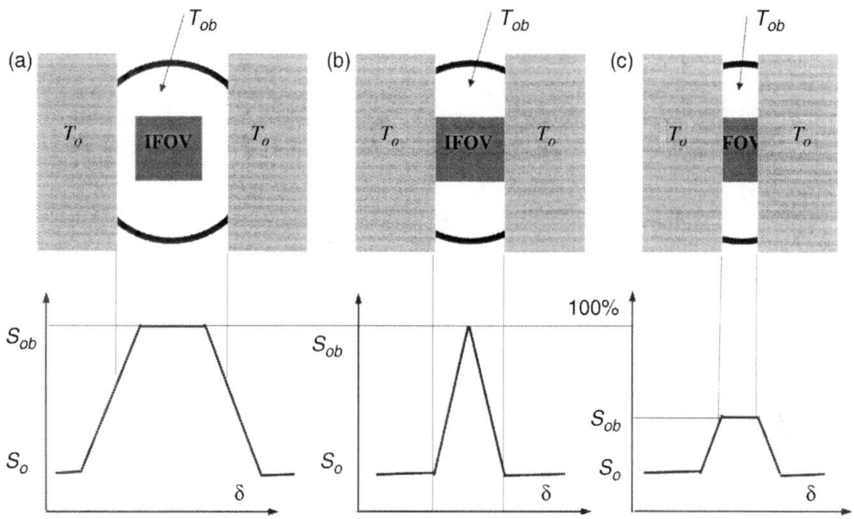

Figure 2.16 Detector's response signal s_{ob} versus observation slit width δ. The size of the object equals IFOV (IR-Book2000). Reproduced by permission of ITC Flir Systems

detector's signal is 10% too small; that is, the indicated temperature is underrated by 10%. It is too big and an unacceptable value. An acceptable value of the modulation function should not be smaller than 98%. In Figure 2.17b it corresponds to angle of flare α_{rad} = 5–8 mrad for measurement cameras (characteristics 1, 2) and α_{rad} = 15–20 mrad for imaging cameras (characteristic 3). The above remarks on the modulation characteristics of the slit response function confirm the earlier conclusion that to guarantee correct temperature measurement

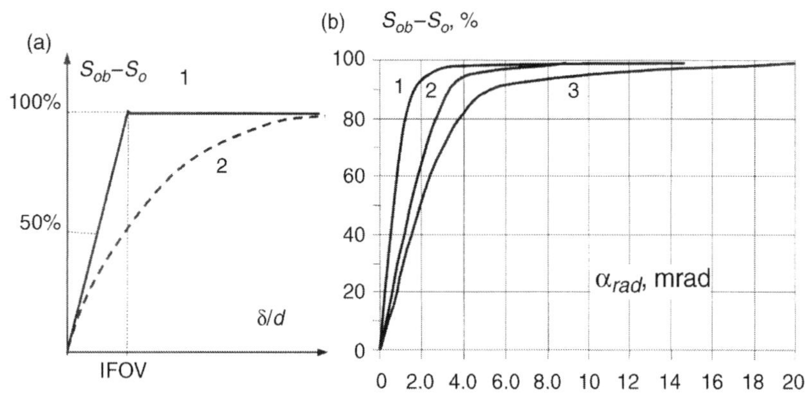

Figure 2.17 Modulation characteristics of difference $s_{ob} - s_o$ versus optical angle of flare of a single detector α_{rad}: (a) generalization: 1, theoretical idealized case; 2, real camera; (b) typical characteristics: 1, 2, measurement camera; 3, imaging camera (Minkina2004). Reproduced by permission of Częstochowa University of Technology

with a camera of a given spatial resolution, determined by the IFOV parameter, the size of an object should not be smaller than 3×3 to 5×5 IFOV.

The described parameters are very important for accurate temperature measurement with an IR camera. Another very important component of the measurement path is the microcontroller processing the measured signals. It realizes a programmed algorithm which, in turn, is worked out on thc basis of a mathematical model of the measurement. In the next chapter we discuss in detail both the algorithm of the measurement path processing and a generalized mathematical model of IR thermography measurement.

3

Algorithm of Infrared Camera Measurement Processing Path

3.1 Information Processing in Measurement Paths of Infrared Cameras

The processing algorithm of an infrared camera is important for the estimation of measurement uncertainty in the thermography method. This algorithm determines how measurement data are obtained from the detector's signals. Signal processing in an infrared camera measurement path can be divided into the following stages (Minkina *et al.* 2003, Minkina and Dudzik 2005):

- detection of infrared radiation in the array detector;
- array calibration or mapping (i.e. linearization and temperature compensation of signals from individual detectors of the array);
- processing of compensated signals by the camera measurement algorithm according to a suitable measurement model.

The first component in the measurement path of a thermograph, regardless of whether it is a scanner or an infrared camera, is an infrared detector.

3.1.1 Infrared Detectors

In the work of Rogalski and colleagues (Bielecki and Rogalski 2001, Rogalski 2000, Rogalski 2003), infrared detectors are divided into thermal and photonic detectors. A division into cooled and non-cooled (operating in the ambient temperature) detectors is often proposed as well. Until 1997 all manufactured infrared cameras were equipped with detectors cooled to temperature from -70 (seldom) to $-200\,^{\circ}$C (most often). Another division, based on the detector's construction, is into single, linear or array (FPA, Focal Plane Array) detectors. Array detectors are matrices of, for example, 640×480 single detectors (pixels) and are standard today.

Infrared Thermography: Errors and Uncertainties Waldemar Minkina and Sebastian Dudzik

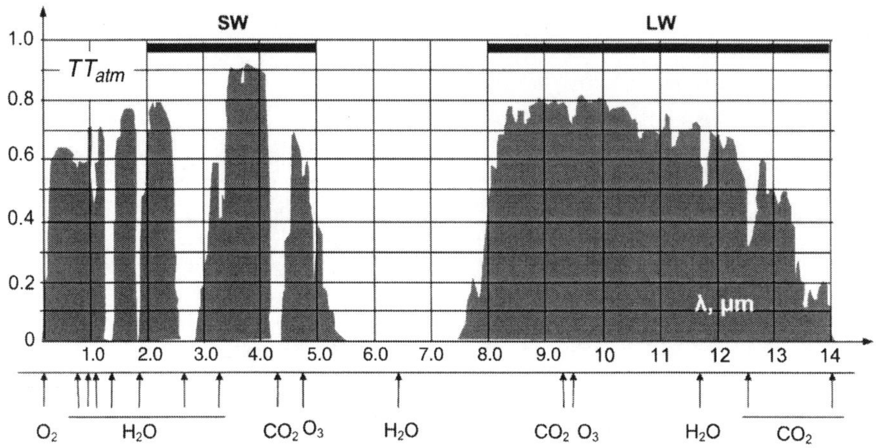

Figure 3.1a Spectral transmission of infrared radiation across a layer of the Earth's atmosphere TT_{atm} of 'thickness' $d = 1.5$ km with absorption bands of the most common gases (Minkina 2004). Reproduced by permission of Częstochowa University of Technology

It is known that atmospheric transmission depends strongly on the radiation wavelength (Figure 3.1) (DeWitt 1983, Schael and Rothe 2002). Therefore, infrared devices operate in bands of maximum possible transmission (minimum absorption). The operating band is yet another criterion of division. There are two infrared bands of best transmission so, traditionally, two types of detectors are distinguished most frequently:

- short-wave (SW) detectors operating in the range 2–5 μm;
- long-wave (LW) detectors operating in the range 8–14 μm.

Another operating band division can be found for instance in Maldague (2001).

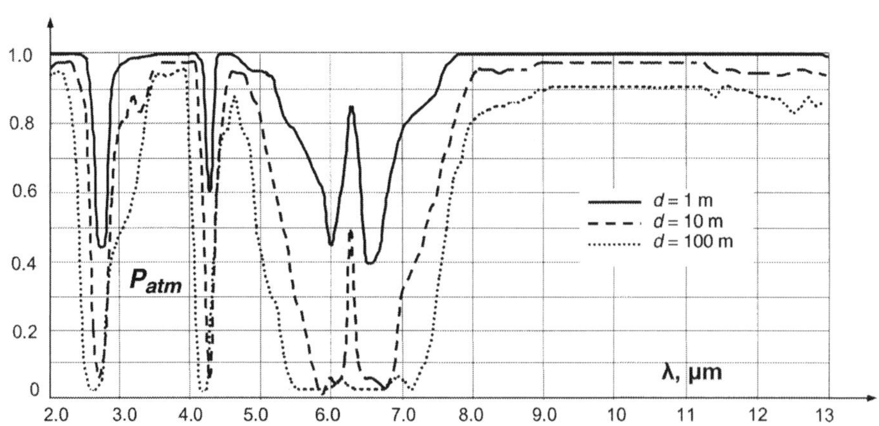

Figure 3.1b Changes in the atmospheric transmissivity TT_{atm} for different values of camera-to-object distance d (Minkina 2004). Reproduced by permission of Częstochowa University of Technology

The manufacture of infrared detectors is currently a rapidly developing branch of technology. Each year hundreds of original publications, surveys and patents are produced on this subject. The basic types of thermal detectors are listed below.

Bolometric detectors are resistors with a very small heat capacity and a big negative temperature coefficient of resistivity. As in the case of thermistors, this coefficient is determined as:

$$\alpha_T = \frac{1}{R_T}\frac{dR_T}{dT} = -\frac{B}{T^2}. \tag{3.1}$$

Measured infrared radiation changes the resistance of bolometric detectors. Metallic bolometers, made of thin foil or evaporated layers of nickel, bismuth or antimony, are still in use today. They can work at room temperature. Semiconductor, superconductor and ferroelectric bolometers are also manufactured.

The structure of a single pixel of a non-cooled array detector (microbolometer) is show in Figure 3.2. The detector absorbs infrared radiation of wavelength $\lambda = 8\text{–}14\,\mu\text{m}$. The microbridge is supported by two metal pins fixed in the silicon base. The pins work also as connectors of the thermometer with the ROIC system. The microbridge contains a thin (0.1 μm) layer of refined synthetic amorphous silicon doped with hydrogen. This layer works

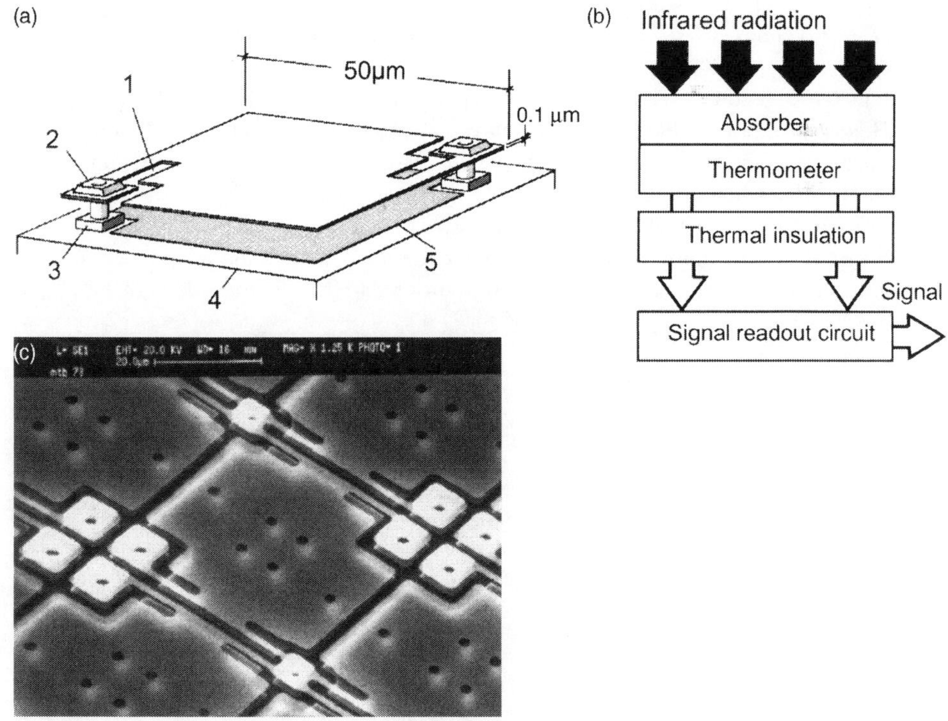

Figure 3.2 (a) Structure of a single pixel: 1, section of thermal insulation; 2, metal pin; 3, metal washer of the ROIC circuit (readout integrated circuit); 4, ROIC circuit; 5, reflecting layer. (b) Block diagram of signal processing. (c) Image from a scanning electron microscope (Tissot *et al.* 1999)

as a thermometer of sensitivity 2.5% K^{-1} that does not significantly absorb the radiation. The radiation is absorbed by a very thin (8 nm), reactively deposited film of titanium nitride. A thermal insulation layer ($\sim 1.2 \cdot 10^7$ K W^{-1}) insulates the thermometer from the information readout circuit. The role of the reflector (aluminum layer) placed on the ROIC surface is to reflect the infrared radiation that penetrated the microbridge back to the thermometer. The size of the single pixel shown in Figure 3.2 is 50 μm. The array detector is a matrix of 256 × 64 pixels. Signal readout is realized by multiplexing each pixel to the ROIC system. The whole readout cycle takes 40 ms. The readout frequency is 25 or 50 Hz for the PAL signal (European standard) and 30 or 60 Hz for the NTSC signal (US standard).

Microbolometric detectors operate at room temperature (i.e. ~300 K), stabilized with a Peltier cooler. Therefore, in contrast to detectors cooled to cryogenic temperatures described further in this chapter, they are called non-cooled detectors. They have been produced since the mid-1990s and currently are very common.

Thermopile detectors are built as a thermopile, that is a system of thermal elements connected in series. A measurement junction is connected to a photosensitive element illuminated by infrared radiation. The temperature of the active surface grows from T to $T + \Delta T$ due to absorbed radiation and the junction becomes heated. The difference of the junction's temperatures generates thermoelectric power:

$$E = k(T_1 - T_o), \tag{3.2}$$

where $(T_1 - T_o)$ is the difference of the junction's temperatures (K) and k the thermoelectric coefficient (μV K^{-1}).

Pyroelectric detectors are built from semiconductors that exhibit the so-called pyroelectric effect. Below the Curie temperature T_C (Figure 3.3) any change in a detector's temperature results in a change in its surface charge, which produces an electric current that can be measured by the ROIC circuit.

The pyroelectric effect is characterized by the so-called pyroelectric coefficient $p(T)$, the ratio of material polarization P and detector temperature T. By adding an electric field, the pyroelectric effect can be increased to a value proportional to (temperature-dependent) electric

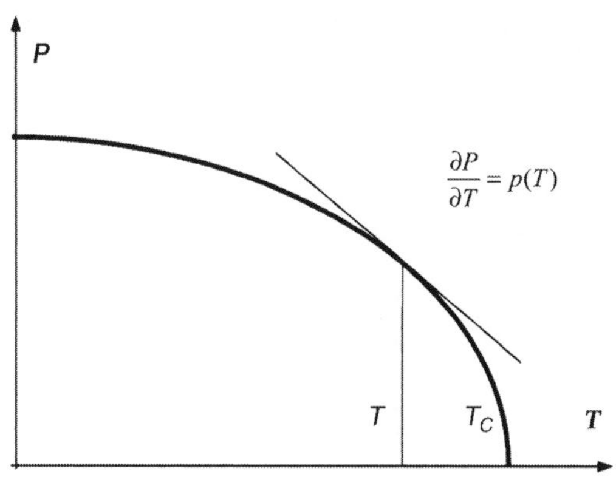

Figure 3.3 Spontaneous polarization P of a pyroelectric detector versus temperature T (Tissot *et al.* 1999). Reproduced by permission of Częstochowa University of Technology

permittivity $\mu(T)$ of the dielectric. This is called the 'field amplification of pyroelectric effect' or 'bolometric ferroelectric effect'. Pyroelectric coefficient p is described as:

$$p = p_o + \int_0^E \frac{\partial \mu}{\partial T} \cdot dE, \tag{3.3}$$

where: p_o is the pyroelectric coefficient without polarization; μ is the absolute electric permittivity (F m^{-1}); and E is the electric field intensity (V m^{-1}).

Pyroelectric detectors are sensitive to rate of temperature change rather than to temperature change itself. It is a property that distinguishes them from other thermal detectors. Therefore, special diaphragms vibrating at a frequency of 25 (30) or 50 (60) Hz have to be applied in infrared cameras with pyroelectric detectors to compare the level of radiation incident on two adjoining detectors (pixels). If a difference in the radiation intensity occurs, a signal representing the difference is generated. Otherwise, the detector does not respond. That is why pyroelectric detectors are sometimes used as motion sensors. They are also manufactured as non-cooled detectors.

Photon detectors are the second type of infrared detectors. They can be divided into the following subtypes.

Photoconductive detectors (photoresistors or photoconductive cells) are detectors with so-called internal photoelectric emission. Infrared radiation falling on a photoresistor changes its resistance. Changes of conductivity are measured on contacts connected to the detector plate. A transverse geometry of photoresistors is usually applied, where incident radiation is perpendicular to the direction of the polarizing current. A change in the voltage drop of a resistor connected in series with the detector is the measured signal. In the case of high-resistance detectors, a constant voltage circuit is preferred. The measured signal is then the current in the detector circuit.

Photovoltaic detectors are detectors also with so-called internal photoelectric emission. They are built from structures containing built-in potential barriers. The photovoltaic effect occurs when redundant carriers are injected near to the barriers. The barriers can be photodiodes with p–n or Schottky junctions.

Photoemissive detectors are detectors with so-called external photoelectric emission. In this case, electrons are ejected from a photocathode material by incident photons and emitted outwards. Photons are absorbed by the photocathode material deposited on a special base which is often transparent to the incident radiation.

Detectors based on quantum wells were developed by AT&T in the early 1990s. Their structure is built from thin foils of AlGaAs and GaAs. To ensure optimum operating conditions they require cooling down to temperature of $-203\,°C$ ($\sim$70 K), more than typical cooled detectors that require $-196\,°C$ ($\sim$77 K). Stirling coolers built in Dewar flasks are used for cooling. Quantum well detectors are currently the most sensitive infrared detectors with a temperature resolution of 20–40 mK. For this reason, they are used mainly in demanding scientific research. They have the best spectral detectivity in a narrow subband (1 μm wide) of the LW band, from 8 to 9 μm. Another characteristic feature of these detectors is the relatively high homogeneity of individual elements (pixels) in the array. Images can be recorded with dynamics corresponding to 14 bit resolution (i.e. $2^{14} = 16\,384$ quantization levels) of an analog-to-digital converter. Information on the basic features of photon detectors is presented in Table 3.1 (Rogalski 2003).

Table 3.1 Basic features of photon detectors

Detector type	Carrier excitation	Electrical signal	Examples of detectors
Intrinsic	Interband	Photoconductivity,	Si, GaN, InSb, HgCdTe
		photovoltage,	Si, InGaAs, InSb, HgCdTe
		capacity	Si, InSb, HgCdTe
Doped	Doping level – band	Photoconductivity	Si:In, Si:Ga, Ge:Cu
Free carrier	Intraband	Photoemission	GaAs/CsO, PtSi
QWIP	Between discrete quantum levels	Photoconductivity, photovoltage	GaAs/AlGaAs InAs/InGaSb

Since detection of the infrared radiation is the first operation carried out by the measurement path of an infrared system, it is important to determine how the detector's properties affect measurement accuracy. This influence is described by appropriate metrologic parameters (Minkina 2004).

3.1.2 Metrologic Parameters of Infrared Detectors

The basic parameters of infrared detectors have been described, for example, in monographs (Rogalski 2000, Rogalski 2003) and in work by Breiter *et al.* (2000), Breiter *et al.* (2002), Tissot *et al.* (1999) and Minkina *et al.* (2000). The parameters that are most important for the processing accuracy of an infrared camera measurement path are discussed below.

Voltage or current (spectral) sensitivity is defined as the ratio of the RMS value of the first harmonic of a detector output voltage (current) to the RMS value of the first harmonic of incident radiation power. For infrared detectors, the spectral sensitivity is given as sensitivity to black body radiation at a defined temperature, usually 500 K.

Temperature sensitivity is a parameter that determines the change of a signal per unit change of temperature for object temperature $T_{ob} = T_o$.

Response rate is a parameter determined by the detector's time constant. Typically, time constants of thermal detectors extend from milliseconds to several seconds (pyroelectric, pneumatic detectors); that is, they are relatively slow. Limit frequencies of the fastest thermal detectors do not exceed several hundred hertz. Photon detectors are much faster. Their limit frequency reaches several hundred megahertz. The time constant of a detector should be as small as possible to enable the recording of thermograms of fast or even ultrafast heat processes (ultrafast thermography is a rapidly developing branch of thermal measurements). Infrared cameras for fast recording are equipped with ultrafast photon detectors.

Noise equivalent power (NEP) is the RMS power of incident monochromatic radiation of wavelength λ that generates the output voltage, whose RMS value is equal to the level of noise normalized to unit bandwidth. In other words, it is the power of radiation required for obtaining the unit signal-to-noise ratio at the detector output. NEP_λ is expressed in W. Since the RMS noise voltage is proportional to the square root of the noise bandwidth, NEP_λ is defined also for a specified bandwidth, usually 1 Hz. 'NEP per unit bandwidth' defined in this way is expressed in W $\mathrm{Hz}^{-1/2}$.

Normalized spectral detectivity D^* is defined as the inverse of NEP_λ. It is the parameter most frequently used for the comparison and assessment of metrological properties of detectors. It is calculated as:

$$D^*(\lambda, f) = \frac{\sqrt{F_d \cdot \Delta f}}{\mathrm{NEP}_\lambda}, \mathrm{cm\,Hz^{1/2}\,W^{-1}}, \tag{3.4}$$

where F_d is the detector's active surface (cm^2) and Δf the frequency bandwidth (Hz).

Detectivity index D^* is related to the detector's unit surface and to unit bandwidth. It is expressed in cm Hz$^{1/2}$ W^{-1} (1 cm Hz$^{1/2}$ W^{-1} is named a 'jones' after R.C. Jones). The normalized detectivity determines the signal-to-noise ratio normalized with respect to the used frequency bandwidth and detector's active surface for incident thermal radiation of unit power. The bigger the detectivity and the broader the bandwidth of the used frequency, the better the detector.

Photos of example array detectors used in infrared cameras produced by FLIR (www.flir.com, www.flir.com.pl, www.infraredtraining.com) and Raytheon (www.raytheon.com) are presented in Figure 3.4.

The next stage of information processing in the measurement path of an infrared camera with an array detector is linearization and temperature compensation of signals from individual detectors (pixels), called 'array calibration' or 'mapping' (Minkina 2004, Nowakowski 2001, IR-Book 2000).

3.1.3 Signal Processing by the Camera Measurement Algorithm

An array detector consists of up to several hundred thousand pixel detectors. Each of them has, in general, different processing characteristics:

$$s_j = f(M_j), \tag{3.5}$$

where s_j is the output signal and M_j the radiation intensity.

The spread of processing characteristics depends on the array type. When a camera is switched on, the detectors are not calibrated. This state is illustrated by an example thermogram

Figure 3.4 Array detectors (FPA) used in cameras from FLIR: (a) non-cooled microbolometer with thermoelectric stabilization (Peltier element), operating temperature about + 30 °C (www.flir.com, www.flir.com.pl, www.infraredtraining.com); and Raytheon (www.raytheon.com): (b) 512 × 512 ALADDIN III Quadrant; (c) 1024 × 1024 ALADDIN III. See Color Plate 5 for the color version

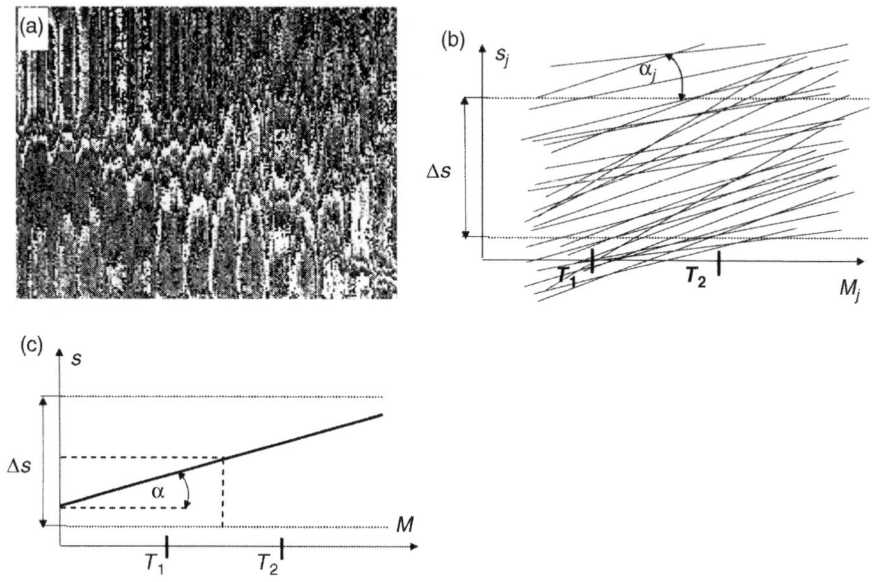

Figure 3.5 Thermogram from an uncalibrated camera and calibration of pixel detector's static processing characteristics $s_j = f(M_j)$ (M_j, radiation exitance at jth detector; s_j, jth detector output signal) (QWIP Seminar 2000). Reproduced by permission of ITC Flir Systems

shown in Figure 3.5a. The recorded thermogram corresponds to a set of static processing characteristics of pixel detectors presented in Figure 3.5b. For correct measurement, the detectors need to be calibrated to the same input/output characteristic (Figure 3.5c). Calibration is performed automatically at each power-on. It proceeds in three steps (IR-Book 2000, Minkina 2004):

- **Step I** – adjustment of vertical ranges of all static characteristics to the same interval Δs (Figure 3.5b), corresponding to the range of the A/D converter used in the camera ROIC circuit. Typically, its resolution is 12 or 16 bits.
- **Step II** – equalization of slope coefficients α_j of the characteristics (Figure 3.5b).
- **Step III** – correction of all static processing characteristics to the same one (Figure 3.5c) so that the midpoint of a given camera measurement range T_1–T_2 corresponds to the midpoint of the A/D converter measurement range Δs.

The camera microcontroller recalculates values of signal s from the calibrated detector array into temperature T according to the measurement model described in Section 3.2. To do that, it is necessary to evaluate calibration constants R, B, F. Evaluation of these calibration constants is made individually for each part of the camera. This procedure is described in Section 4.2.

To evaluate the temperature of an observed object, the microcontroller performs temperature compensation of each pixel. It is necessary to eliminate the influence of thermal self-radiation of the camera, the detector's reference temperature and linearization of the processing characteristics described above (Figure 3.5). The compensation is carried out using the following formula (TOOLKIT IC2):

$$absPixel = globalGain \cdot LFunc(imgPixel) + globalOffset, \quad (3.6)$$

where: *absPixel* is the value of a pixel after compensation; *LFunc* is the value of a processing characteristic linearization function for argument *imgPixel*, which is the value of a pixel before compensation ('raw' pixel); *globalGain* and *globalOffset* are constants, corresponding to parameters of instrumentation amplifiers of the measurement path.

Linearizing function *LFunc* is based on two coefficients: *Obas*, the base offset used in nonlinear conversion; and *L*, the calibration constant linearizing raw pixel values. Function *LFunc* is defined as:

$$LFunc = \frac{p - Obas}{1 - L \cdot (p - Obas)}, \quad (3.7)$$

where p is the raw pixel value.

Another important task of the processing algorithm, apart from the evaluation of temperature for each pixel, is to display a recorded thermogram as a color image. To obtain an image of the temperature field, the camera executes an imaging procedure that assigns to temperature readouts values defined by a color map (color palette). The imaging algorithm has to conduct such a color assignment both in the camera and in the thermogram processing software on a PC. Below we describe the color imaging algorithm applied in the original TermoLab software, working with cameras produced by FLIR.

TermoLab is a thermal analysis system (Minkina *et al.* 2002, Minkina *et al.* 2003), whose algorithm uses radiometric data stored in a thermogram file (*.img), created by the camera microcontroller in a special format (AFF, AGEMA File Format) elaborated by AGEMA/FLIR. An AFF file includes a table of 16 values that describe the distribution of colors in a thermogram. Two methods of colorizing are shown in Figures 3.6 and 3.7. The first one – isothermal algorithm – is used internally in the camera, the second one – histogram algorithm – is used by the TermoLab software (TOOLKIT IC2, Dudzik 2000).

The histogram algorithm is based on a table stored in the *.img file. This table defines the assignment of values within 16 consecutive intervals determined by the algorithm. The number of colors per interval is calculated after transformation carried out on a table readout from a thermogram file. Next, each pixel is filtered using an assignment algorithm, which allows calculation of the corresponding index in the color map (colorizing table). The final step is

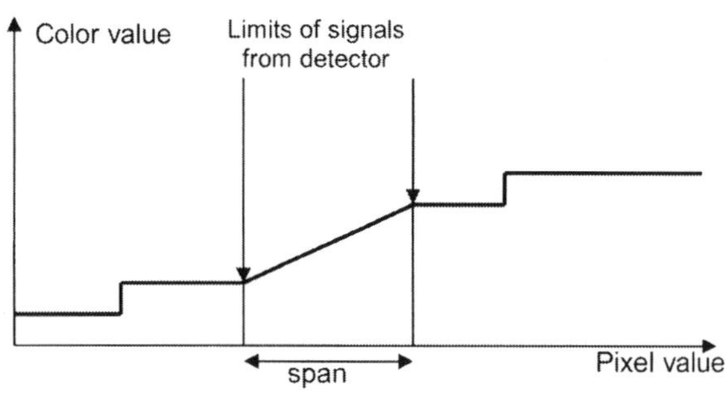

Figure 3.6 Isothermal colorizing algorithm used in FLIR cameras (TOOLKIT IC2)

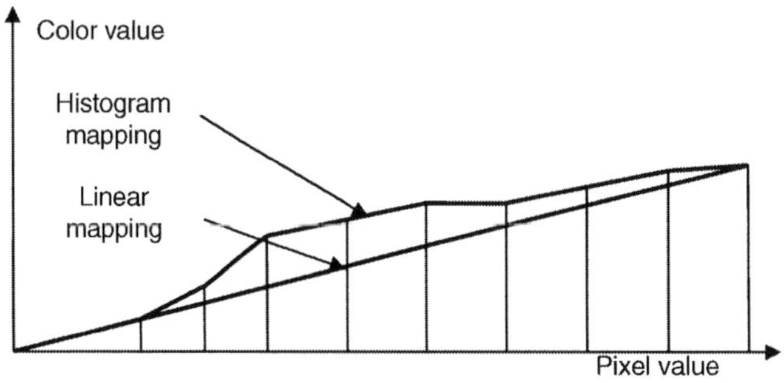

Figure 3.7 Histogram colorizing algorithm used in FLIR cameras (TOOLKIT IC2)

reading a value corresponding to the calculated index from the color map. As a result, each pixel value is represented by color defined in the used color map. The pixels create a thermogram image, displayed on the camera or computer monitor. Examples of enlarged images obtained from a 4 × 4 array detector for different color maps are presented in Figure 3.8. At a smaller real scale the pixels give the impression of a continuous image. Suitable selection of color maps can generate lots of possibilities. For instance, it is possible to create images representing invisible bands of radiation (e.g. ultraviolet, infrared, X-rays, etc.), but we should keep in mind that the colors are conventional. This method of imaging is called 'pseudocolorizing', because selected sets of colors are not associated in reality with measured values or perceived by a human.

The method of pseudocolorizing can be explained by the example of the RGB (Red, Green, Blue) color map. Intermediate colors are produced as mixtures of the three basic colors with

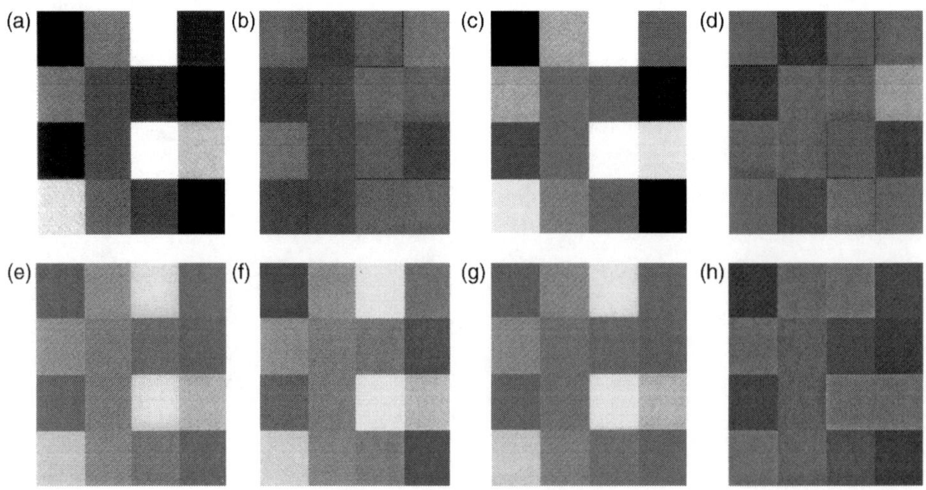

Figure 3.8 Enlarged digital image for different color maps: (a) grayscale; (b) cool; (c) hot; (d) HSI (Hue Saturation Intensity); (e) spring; (f) summer; (g) autumn; (h) winter. See Color Plate 6 for the color version

appropriate weights. That is, color K is obtained as:

$$K = r \cdot R + g \cdot G + b \cdot B, \qquad (3.8)$$

where r, g, b are the weights of the basic colors R, G, B.

To represent an image using the *RGB* components, three matrices with stored weights r, g, b are required. Each weight as well as their sum can have a value from 0 to 1. In computer systems it is easier to store integer values, so the weights are represented by 1 byte (i.e. 8 bits). This allows the representation of $2^8 = 256$ shades of each color component. In general, the 3 byte, 24 bit RGB format allows the representation of $(2^8)^3 = 256 \times 256 \times 256 = 16\,777\,216$ colors. There are also extended RGB formats using 32 bits for color coding. In the 8 bit grayscale, the weights of the color components are equal, $r = g = b$, so it is possible to represent 256 shades of gray.

Most software for thermogram processing, like the TermoLab package above, allows user color maps to be defined (Dudzik 2007). However, the choice of colors must take into consideration that some colors have an additional meaning: for example, red is used for warning. We should emphasize again the distinction between a thermogram – a matrix of data from an array detector – and an image – a graphic presentation of data using a color map or grayscale.

Apart from the grayscale, the following precisely defined color maps are most often used in thermography: rainbow, rainbow 10, iron and iron 10. Number 10 denotes that the software displays a thermogram for a given color map using only 10 color shades. Other color maps, like glowbow, grayred, medical, midgreen, midgray or yellow, are used more seldom.

An alternative form of presentation of a thermogram is a 3D graph. The height of a bar in the third dimension is proportional to the temperature of the corresponding pixel. The equivalent of a pixel on a 3D thermogram is called a **voxel** (volumetric pixel). The 3D presentation can be more useful for the qualitative evaluation of heat processes. The additional dimension can also represent time. It is helpful when temperature changes rapidly and it is difficult to investigate the changes on a numerous set of thermograms recorded in a sequence of measurements.

The size of a thermogram (in pixels) can be bigger than the original size of a detector array thanks to either interpolation that generates a bigger matrix (Nowakowski 2001) or subpixel processing taking into consideration the model of near-edge temperature damming (Astarita *et al.* 2000, Bayazitoğlu and Özişik 1998, Bąbka and Minkina 2001, Minkina *et al.* 2001, Bąbka and Minkina 2002a,b, Bąbka and Minkina 2003a). Detailed discussion of digital thermogram processing goes beyond the scope of this book. More information on this subject can be found in Nowakowski (2001), for example.

The final operation of an infrared system measurement path algorithm is the processing of the compensated signal into a value of the surface temperature. The basis of this processing is the mathematical model of measurement with an infrared camera (Minkina 2004, Minkina and Dudzik 2005, Minkina and Dudzik 2006a). The accuracy of temperature evaluation with an infrared system depends strongly on the error in the method, conditioned by the applied measurement model. The next section is devoted to this issue.

3.2 Mathematical Model of Measurement with Infrared Camera

The foundation of temperature measurement with an infrared camera is the theory of thermal radiation of bodies (Schuster and Kolobrodov 2000, Stahl and Miosga 1986, Walther and

Gerber 1983). In a mathematical model of temperature measurement it is necessary to take into account the following heat fluxes arriving at an infrared detector (Minkina 2004, Minkina and Bąbka 2002):

- flux φ_{ob} emitted by the investigated object;
- flux φ_{refl} emitted by the ambient and reflected from the investigated object;
- flux φ_{atm} emitted by the atmosphere;
- flux emitted by optical components and filters of the camera; in the most recent models its influence on measurement is considered negligible.

These fluxes can be expressed as:

$$\varphi_{ob} = \varepsilon_{ob}(T_{ob})TT_{atm}(T_{atm})M_{ob}(T_{ob}), \tag{3.9a}$$

$$\varphi_{refl} = [1 - \varepsilon_{ob}(T_o)]\varepsilon_o(T_o)TT_{atm}(T_{atm})M_o(T_o), \tag{3.9b}$$

$$\varphi_{atm} = [1 - P_{atm}(T_{atm})]M_{atm}(T_{atm}), \tag{3.9c}$$

where: ε_{ob} is the band emissivity of the object surface; ε_o is the band emissivity of the ambient; M_{atm}, M_{ob}, M_o are the radiant exitance of the atmosphere, object and ambient, respectively (W m^{-2}); TT_{atm} is the band transmittance of the atmosphere; T_{atm}, T_{ob}, T_o are the temperature of the atmosphere, object and ambient, respectively (K).

A diagram illustrating the interaction of the heat fluxes is shown in Figure 3.9.

The output signal from the camera detector can be described by the formula:

$$s \approx C(\varphi_{ob} + \varphi_{odb} + \varphi_{atm}), \tag{3.10}$$

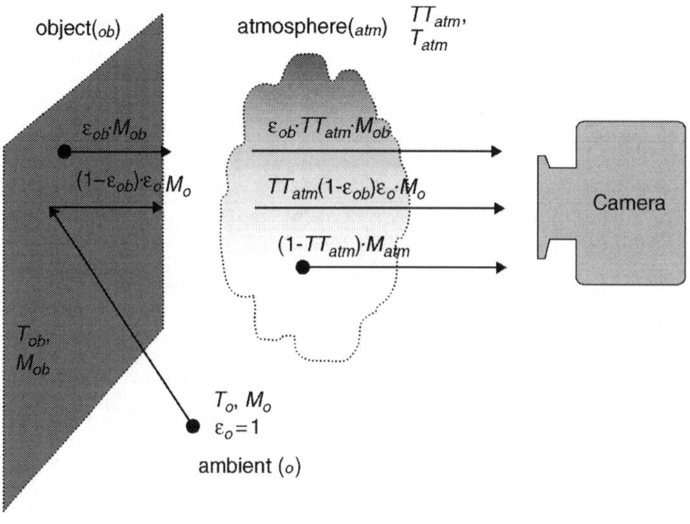

Figure 3.9 Interaction of the radiation fluxes in measurement with an infrared camera (Minkina 2004, IR-Book 2000). Reproduced by permission of Częstochowa University of Technology

where C is a parameter depending on atmospheric damping, the camera's optical components and the detector's properties.

On the basis of (3.9) and (3.10), the measurement model can be expressed as (Minkina and Dudzik 2005):

$$s = \varepsilon_{ob} \cdot TT_{atm} s_{ob} + TT_{atm} \cdot (1 - \varepsilon_{ob}) s_o + (1 - TT_{atm}) s_o, \tag{3.11}$$

where: s is the detector signal corresponding to the total intensity of heat radiation arriving at the detector; s_{ob} and s_o are the detector signals corresponding to the object heat radiation intensity and a black body heat radiation intensity, at ambient temperature, respectively.

Signal s_o is expressed as:

$$s_o = \frac{R}{\exp(B/T_o) - F}, \tag{3.12}$$

where R, B, F are constants associated with the camera calibration characteristic, described further in Section 4.2.

Taking into account (3.11) and (3.12), we can derive the detector signal corresponding to the radiation flux density of the investigated object as:

$$s_{ob} = s \frac{1}{\varepsilon_{ob} TT_{atm}} - \left[\frac{1 - \varepsilon_{ob}}{\varepsilon_{ob}} \frac{R}{\exp(B/T_o) - F} + \frac{1 - TT_{atm}}{\varepsilon_{ob} \cdot TT_{atm}} \frac{R}{\exp(B/T_o) - F} \right] \tag{3.13}$$

Coefficient TT_{atm}, associated with absorption of the infrared radiation by a layer of the atmosphere, plays an important role in this equation. The absorption is, in turn, caused by molecules of steam (H_2O), carbon dioxide (CO_2) and ozone (O_3). The concentration of these compounds in the atmosphere varies with weather, climate, season or geographic location. As mentioned in Section 3.1, there are bands of smaller absorption of infrared radiation, called atmospheric windows, that enable infrared thermography measurements: the SW window (2–5 μm, atmospheric window I) and the LW window (8–14 μm, atmospheric window II). Consequently, infrared cameras are divided in a natural way into SW and LW cameras.

It has been observed, even under laboratory conditions, that at a distance of 1–10 m, the atmospheric absorption, caused by steam and carbon dioxide, is noticeable. It is depicted in Figure 3.1. The most important role in absorption of the infrared radiation for wavelength $\lambda = 4.3$ μm is played by carbon dioxide, present in exhaled air. Rudowski (1978) stated, for example, that after 3 hours in a closed room of about 40 cubic meters in volume, the concentration of CO_2 exhaled by two people was such that at a distance $d = 0.8$ m, 70% of radiation of wavelength $\lambda = 4.3$ μm was absorbed by the air.

Depending on the infrared camera model, there are several different models of atmospheric transmission, such as FASCODE, MITRAN, MODTRAN, PcModWin, SENTRAN, WATRA and others (Pręgowski and Świderski 1996, Pręgowski 2001). For example, in the AGEMA 470 Pro SW and AGEMA 880 LW systems, the manufacturer employs the following formula for the LOWTRAN transmission model:

$$TT_{atm}(d) = \exp\left[-\alpha\left(\sqrt{d} - \sqrt{d_{cal}}\right) - \beta(d - d_{cal}) \right], \tag{3.14}$$

with the following values for the coefficients:

- for an SW camera: $\alpha = 0.003\,93\ \text{m}^{-1/2}$, $\beta = 0.000\,49\ \text{m}^{-1}$;
- for an LW camera: $\alpha = 0.008\ \text{m}^{-1/2}$, $\beta = 0\ \text{m}^{-1}$.

The given values are determined under the normal conditions of atmospheric temperature $T_{atm} = 15\,°\text{C}$ and relative humidity $\omega_{\%} = 50\%$. Under different conditions, the atmospheric transmittance model will be different. The value of coefficient TT_{atm} versus distance d between the camera and the object is shown in Figure 3.10a for an LW (1) and SW camera (2). These relationships were obtained from numerical computations using formula (3.14). One can see that the atmosphere has better transmission within the LW infrared band. Very similar results are presented in Pręgowski (2001).

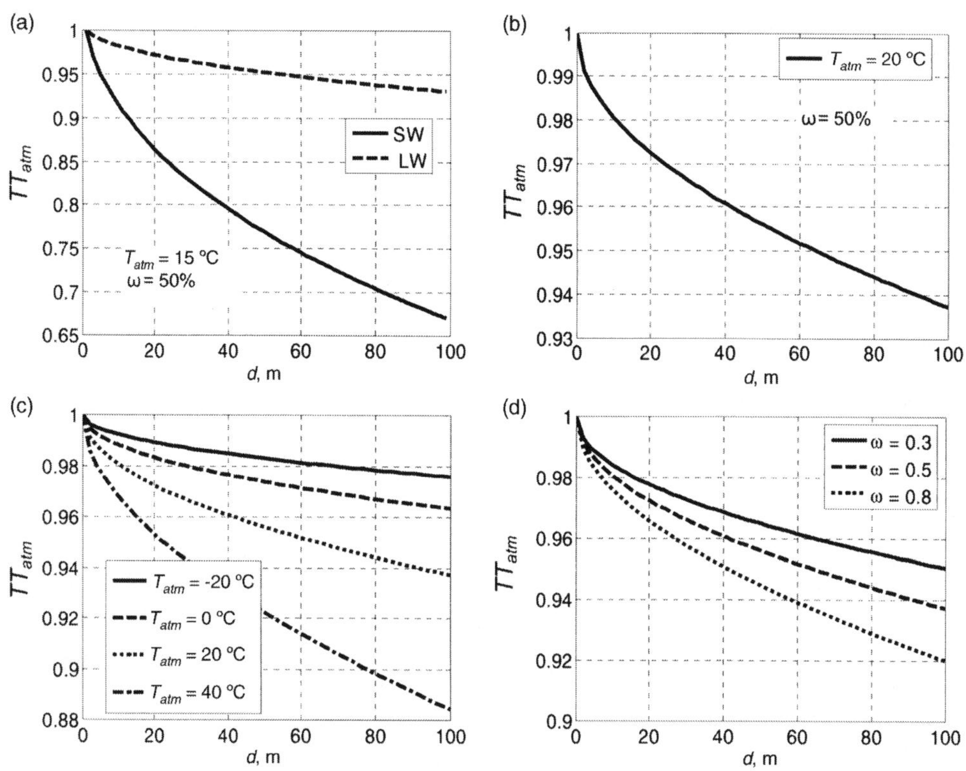

Figure 3.10 Simulation characteristics of the atmospheric transmission TT_{atm} versus camera-to-object distance d for different transmission models: (a) $T_{atm} = 15\,°\text{C}$, $\omega = 50\%$, LOWTRAN model (3.14) for LW and SW camera; and for comparison: (b) $T_{atm} = 20\,°\text{C}$, $\omega = 50\%$ model (3.15) for ThermaCAM PM 595 LW camera; (c) $\omega = 50\%$, model (3.15) for ThermaCAM PM 595 LW camera; (d) $T_{atm} = 20\,°\text{C}$, model (3.15) for ThermaCAM PM 595 LW camera

The transmittance model defined by FLIR for the ThermaCAM PM 595 camera is a function of three variables: atmospheric relative humidity $\omega_{\%}$, camera-to-object distance d and atmospheric temperature T_{atm} (Minkina and Dudzik 2005, Dudzik 2007):

$$TT_{atm} = f(\omega_{\%}, d, T_{atm}) \tag{3.15}$$

This model will be applied to error and uncertainty analysis performed later in this monograph. It is actually *very complex*. It includes, among others, nine coefficients adjusted empirically. The explicit form of function (3.15) is proprietary and reserved by the camera manufacturer (TOOLKIT IC2). It was made available to us only for research purposes, so we cannot publish it here. We are allowed to present the characteristic of the atmospheric transmission TT_{atm} as a function of camera-to-object distance d. The results, shown in Figures 3.10b–d, were obtained by numerical simulations using the full form of (3.15).

It must be emphasized that the model described by formula (3.15) concerns most infrared cameras produced by the AGEMA Company (e.g. 900 series) and FLIR Company (e.g. ThermaCAM PM 595 LW).

The example experimental characteristics of transmission TT_{atm} versus distance d for an LW and SW camera are presented in Figure 3.11. The measurements were carried out for two different values of relative humidity $\omega_{\%}$ and three values of object temperature T_{ob}. The goal of the experiment was to check how the object temperature affects its 'visibility' in the infrared under given conditions.

Taking into account (3.12)–(3.15), the object temperature can be expressed as:

$$T_{ob} = \frac{B}{\ln\left(\dfrac{R}{s_{ob}} + F\right)}, \text{K}. \tag{3.16}$$

Finally, the infrared camera measurement model is defined as a function of five variables:

$$T_{ob} = f(\varepsilon_{ob}, T_{atm}, T_o, \omega, d), \text{K}. \tag{3.17}$$

We must emphasize that the model derived above is a simplified model. In reality, the camera detector receives radiation not only from the object, but also from other sources. The simplification can be explained by looking at Figure 3.12. Signal s_o appearing in formula (3.12), proportional to the ambient radiation intensity and dependent on ambient temperature T_o (in (3.12)), is in reality an average response to radiation coming from clouds of temperature T_{cl}, from buildings of temperature T_b, from ground of temperature T_{gr} and from the atmospheric temperature T_{atm}. All these temperatures differ a little from each other (Orlove 1982, DeWitt 1983, Saunders 1999).

The above description of phenomena constituting measurements in infrared thermography does not encompass all possible measurement situations. For example, an investigated object can be located in a furnace chamber, in a vacuum chamber or in a wind tunnel. In such cases, the camera looks through an inspection window. The material for this window must be transparent within the camera's operational bandwidth (SW: 2–5 µm; LW: 8–14 µm). To make possible visual control of the measurement, the window should also be transparent within the visible

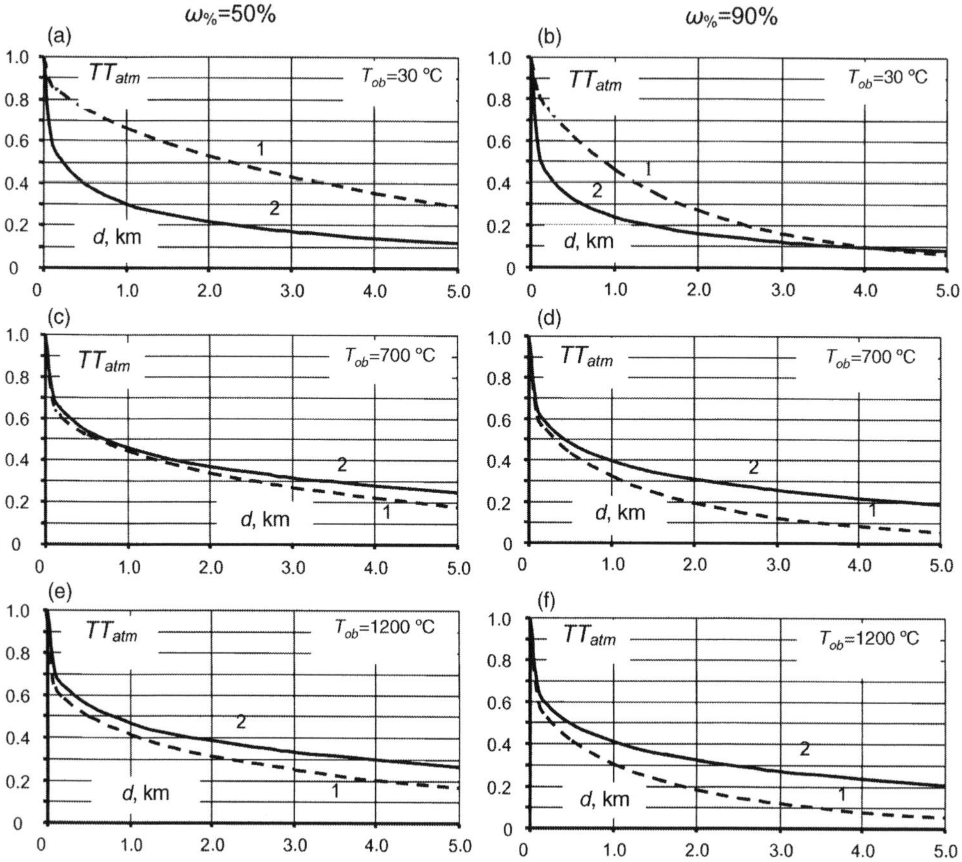

Figure 3.11 Examples of object 'visibility' characteristics taking into consideration the atmospheric transmission TT_{atm}, camera-to-object distance d, atmospheric relative humidity $\omega_{\%}$ and camera type: 1, LW camera; 2, SW camera (IR-Book 2000). Reproduced by permission of ITC Flir Systems

radiation bandwidth. Other requirements can concern mechanical strength (in case of a difference of pressure inside and outside the window) or resistance to chemicals or to rapid temperature changes. Spectral characteristics of the transmission coefficient of typical materials used for inspection windows are shown in Figure 3.13.

The bands of high transmission shown in Figure 3.13 can widen or narrow with temperature and thickness, depending on material. Manufacturers of inspection windows usually specify window properties at room temperature and for a certain thickness. Information on the influence of temperature and window thickness on the spectral characteristic is not always specified, as well as restrictions concerning mechanical strength and maximum work temperature of a window.

In measurement through an inspection window, the measurement model described above needs to be generalized by taking into account additional fluxes of radiation. Paths of the radiation fluxes in such a case are illustrated in Figure 3.14.

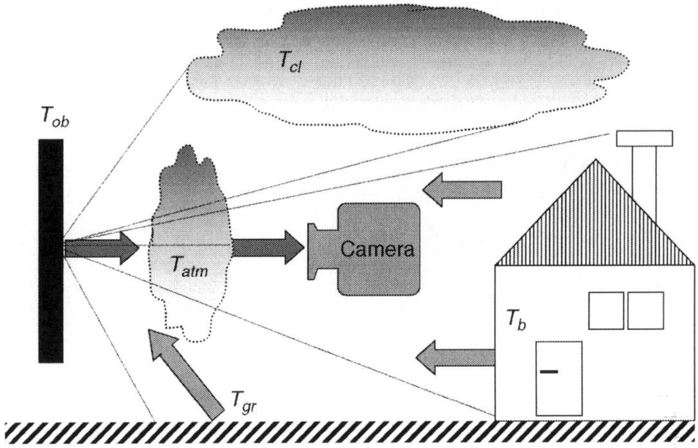

Figure 3.12 Explanation of simplifications assumed in the infrared camera measurement model 3.17; ambient temperature T_o is an average of temperatures of clouds T_{cl}, atmosphere T_{atm}, ground T_{gr} and buildings T_b (Minkina 2004). Reproduced by permission of Częstochowa University of Technology

Taking into account additional fluxes of radiation, the signal arriving at the camera detector can be expressed in the considered model as:

$$\begin{aligned} s = {} & \varepsilon_{ob}(T_{ob}) \cdot TT_{atm1}(T_{atm1}) \cdot TT_w(T_w) \cdot TT_{atm2}(T_{atm2}) \cdot s_{ob}(T_{ob}) \\ & + [1 - \varepsilon_{ob}(T_{o1})] \cdot TT_{atm1}(T_{atm1}) \cdot TT_w(T_w) \cdot TT_{atm2}(T_{atm2}) \cdot s_{o1}(T_{o1}) \\ & + [1 - TT_{atm1}(T_{atm1})] \cdot TT_w(T_w) \cdot TT_{atm2}(T_{atm2}) \cdot s_{atm1}(T_{atm1}) \\ & + \varepsilon_w(T_w) \cdot TT_{atm2}(T_{atm2}) \cdot s_w(T_w) \\ & + R_w(T_{o2}) \cdot TT_{atm2}(T_{atm2}) \cdot s_{o2}(T_{o2}) \\ & + [1 - TT_{atm2}(T_{atm2})] \cdot s_{atm2}(T_{atm2}). \end{aligned} \tag{3.18}$$

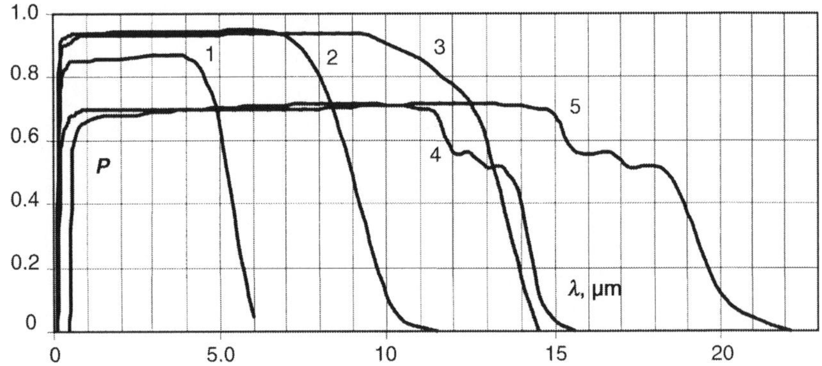

Figure 3.13 Transmission coefficient of several materials used for inspection windows: 1, Al_2O_3; 2, CaF_2; 3, BaF_2; 4, ZnS; 5, ZnSe (IR-Book 2000). Reproduced by permission of ITC Flir Systems

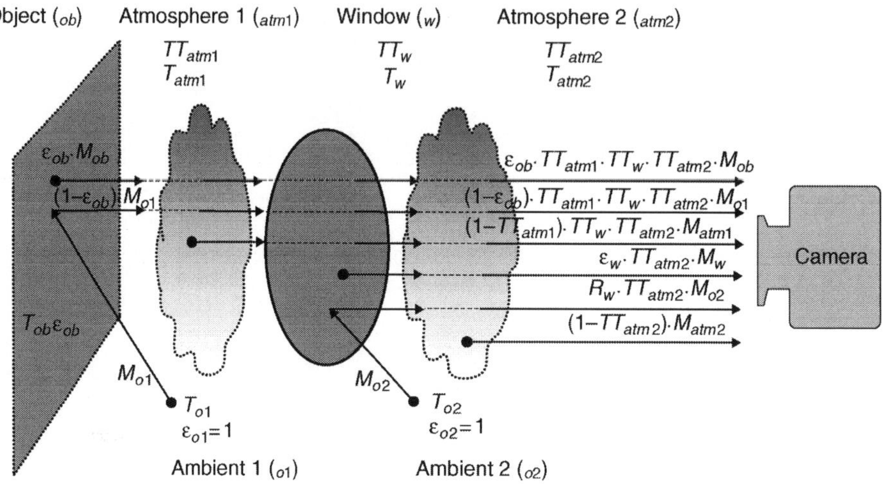

Figure 3.14 Paths of radiation fluxes in measurement through an inspection window (IR-Book 2000). Reproduced by permission of ITC Flir Systems

Using formula (3.18) we can derive s_{ob} (Minkina 2004):

$$\begin{aligned} s_{ob} &= \frac{s}{\varepsilon_{ob} \cdot TT_{atm1} \cdot TT_w \cdot TT_{atm2}} - \frac{(1-\varepsilon_{ob}) \cdot s_{o1}}{\varepsilon_{ob}} - \frac{(1-TT_{atm1}) \cdot s_{atm1}}{\varepsilon_{ob} \cdot TT_{atm1}} \\ &\quad - \frac{\varepsilon_w \cdot s_w}{\varepsilon_{ob} \cdot TT_{atm1} \cdot TT_w} - \frac{R_w \cdot s_{o2}}{\varepsilon_{ob} \cdot TT_{atm1} \cdot TT_w} - \frac{(1-TT_{atm2}) \cdot s_{atm2}}{\varepsilon_{ob} \cdot TT_{atm1} \cdot TT_w \cdot TT_{atm2}}. \end{aligned} \tag{3.19}$$

The measurement model of a typical infrared camera does not take account of all heat fluxes. In order to make a correct measurement it should be applied to the model (3.19). In practice it is hard to allow for all heat fluxes, so the following recommendations should be considered:

- The inspection window should be made from a material which does not absorb radiation in the spectral range of the infrared camera: $TT_w \approx 1$.
- If a typical infrared camera has to be used with the model (3.13), the following issues should be considered:
  - application of the filter with a range in which the atmosphere between the camera and the object has a very good transmission (the best is a vacuum); or
  - location of the camera possibly close to the inspection window (i.e. $TT_{atm2} \approx 1$). In this case the value of TT_{atm1} is entered manually into the camera software.

The way of reasoning used to create the model from Figure 3.14 can also be applied to create an accurate, general measurement model of an infrared camera by taking into consideration radiation from individual optical components and filters (Hamrelius 1991).

The processing algorithm of an infrared camera measurement path operates on the basis of the mathematical model described above. However, this does not mean that the algorithm must be executed on-line by the camera microcontroller. It is very often implemented off-line, and executed by software installed on a PC. In such a case, raw radiometric data (i.e. uncompensated values of pixels, calibration parameters, etc.) are transferred to the computer in files of a special format.

Digital systems of off-line infrared data processing implemented on PCs are usually designed for specific camera types for the sake of different detectors, optical systems and various processing algorithms of recorded data used by different manufacturers. Below we present a brief description of the TermoLab system, designed by one of the authors for his own needs (Dudzik 2007).

The TermoLab system represents a more universal approach and allows for the analysis of temperature fields recorded with a camera of any type. The only requirement imposed on a recording device is the matrix output data format (matrix of temperatures). TermoLab uses the MATLAB format for its input data. Such an approach extends the area of possible system applications. Dedicated manufacturer software usually allows for standard engineering analysis (calculation of temperatures, presentation), basic statistical analysis (mean, maximum, etc.) and creating reports for measurement documentation. TermoLab is designed for research purposes and makes possible advanced statistical analysis of recorded thermograms. A novelty in comparison to typical manufacturers' software is the use of advanced digital image processing (filtering, noise reduction, etc.) and comparative analysis of thermograms with the detection of discrepancy areas (for diagnostic applications). It is possible to extend the system depending on the user's individual needs. These are the basic features of the TermoLab software:

- Compensation of influence due to the object emissivity, ambient and atmospheric radiation. This function is performed by procedures reading additional data (emissivity, air humidity, etc.) from AFF files.
- Compensation of the camera components' self-radiation. The software calibrates the detector readings using coefficients compensating radiation of the camera components.
- Thermogram filtering – imaging of temperature fields using various color maps and achieving optimum matching of the color map to the temperature range.
- Determination of isothermal areas.
- Evaluation of histograms – the system allows a temperature occurrence frequency histogram to be calculated for any thermogram subarea.
- Evaluation of minimum, maximum and mean temperatures for any thermogram subarea.
- Evaluation of temperature distribution profile in any direction – evaluation and analysis of horizontal and vertical profiles are possible.
- Superposition of thermograms – correlation analysis with determination of discrepancy areas.
- Creation of pseudo 3D thermograms.
- Presentation of series of images – simultaneous display and analysis of multiple thermograms thanks to the applied multidocument interface.
- Automation of decision process.

To ensure exchange of information between the camera microcontroller software and dedicated PC software, special communication interfaces and data formats are used. These interfaces and data formats are closely associated with particular camera types. Since cameras available on the market differ significantly from each other, data analysis is most often possible using software from only one specified manufacturer.

TermoLab has a built-in UMI (Universal Matrix Interface) that reads data either as matrices of temperature values stored in MATLAB format files (from any camera type) or as direct detector signal values from AFF files (only from FLIR cameras). The first option enables analysis of data from any camera that stores images as matrices. The system offers extensive possibilities for the statistical analysis of infrared data as well as graphical presentation of results. The reports including results of measurements and computations, generated during program execution, can also be stored in MATLAB format files and further processed in the MATLAB environment if necessary. The AFF format of the input data files is an original product of FLIR (TOOLKIT IC2) and directly supports FLIR (formerly AGEMA) cameras.

So far we have concentrated on describing infrared thermography measurements from the point of view of the measurement method, physical phenomena occurring during measurements and signal processing of the infrared camera measurement path. In the next chapters, we will explore issues concerning the accuracy of infrared thermography in temperature measurements.

In our opinion, error analysis and uncertainty analysis do not exclude each other. On the contrary, taking into consideration the complexity of the measurement in infrared thermography, they complement each other and allow the reader to form a more general opinion of the accuracy of measurement methods in infrared thermography. In this book the errors and uncertainties were calculated for the exemplary infrared FLIR camera ThermaCAM PM 595 LW. For other camera types and manufacturers, the results and conclusions can be *very similar*.

4

Errors of Measurements in Infrared Thermography

4.1 Introduction

The concept of 'measurement error' is of basic significance for the assessment of measurement method accuracy. The evaluation of errors is also necessary in infrared thermography measurements, especially as final values of measured temperature are obtained at the output of a complex processing algorithm (based on a mathematical model of measurement) of an infrared camera measurement path (Minkina 2004). For a correct assessment of accuracy, it is necessary to evaluate the errors in the method introduced by the applied model.

In this monograph we use the definitions of infrared system measurement errors below.

The **absolute error of a measurement model in infrared thermography** is the difference between value T_C calculated by the camera measurement path algorithm for a single element (pixel) of the array detector and actual temperature T_R of the surface area mapped (represented) by this element (Minkina 2004):

$$\Delta_{T_{ob}} = T_C - T_R, \mathrm{K}, \tag{4.1}$$

where T_C is the value of temperature calculated for a single pixel of a thermogram and T_R the actual value of temperature. It is assumed that the temperature of a surface area mapped by a single pixel is constant.

The **relative error of a measurement model in infrared thermography** is the ratio of absolute error $\Delta_{T_{ob}}$ to actual temperature T_R:

$$\delta_{T_{ob}} = \frac{\Delta_{T_{ob}}}{T_R}. \tag{4.2}$$

Unfortunately, the actual value T_R in the above definitions is unknown. Therefore in this monograph we substitute it by a true conventional value, assigned a priori during simulation of the camera processing algorithm.

In Chapter 1 we divided measurement errors into systematic and random errors. In the case of measurements in infrared thermography, systematic interactions strongly affect the accuracy of

Infrared Thermography: Errors and Uncertainties Waldemar Minkina and Sebastian Dudzik

temperature evaluation. Therefore, in the next section we discuss sources of systematic interactions in infrared thermography measurements.

4.2 Systematic Interactions in Infrared Thermography Measurements

In work by Chrzanowski (2000), Minkina *et al.* (2000) and Minkina and Wild (2004), errors of temperature measurement with infrared cameras are classified in the following way:

- errors of method;
- calibration errors;
- electronic path errors.

Under real conditions, errors in the method can follow from the reasons below or interactions occurring during measurement:

- incorrect evaluation of object emissivity ε_{ob} and/or incorrect evaluation of T_{atm}, T_o, ω, d;
- influence of the ambient radiation – direct and/or reflected from the object – arriving at the camera detector;
- incorrect evaluation of atmospheric transmission and atmospheric radiance;
- detector noise.

The more the number of objects of different emissivity imaged in one thermogram, the more noticeable the influence of the emissivity evaluation error. Nevertheless, in modern infrared systems, it is possible to specify the emissivity separately for defined areas of a thermogram in off-line analysis (Dudzik 2000, Dudzik 2003). It reduces the influence of the heterogeneity of the observed surface. As described in Section 2.3, the object emissivity depends on wavelength λ, temperature T, material, state of the surface, direction of observation, polarization and also – in ultrafast thermal processes – on time. Consequently, complete elimination of the influence of incorrect emissivity evaluation seems to be impossible. However, this influence can be significantly reduced by painting the surface black, moistening it or – if possible – by uniform heating and then making a map of the emissivity. Application of these solutions is easy under laboratory conditions but usually impossible under industrial conditions. Therefore, correct evaluation of the measurement error component due to incorrect emissivity is very important. This is more so because object emissivity is one of the input quantities in the camera measurement path algorithm.

The influence of radiation emitted by the ambient grows when ε_{ob} decreases. This follows from (3.11). The effect is more significant for $T_o \geq T_{ob}$. Additional errors due to the Sun's radiation appear in infrared thermography measurements performed outdoors. The Sun can be considered as a black body at high temperature. The Sun's radiation incident on an object is filtered by the atmosphere and depends on day, time and atmospheric conditions. An investigation of the influence of the Sun's radiation on accuracy in infrared thermography is not easy: usually this radiation makes measurement impossible, except during qualitative investigations of high-temperature objects whose $\varepsilon_{ob} \approx 1$. The situation is even more complicated when the observed object reflects sky radiation, buildings radiation and ground radiation (see Figure 3.12). Limitation of the influence of atmospheric radiation can be achieved by entering the ambient temperature into the camera's microcontroller. Unfortunately, the problem is to determine this temperature in a reliable way. It is difficult because the neighborhood of an observed object can encompass many components of various values of emissivity located close to it or farther away. A method of reducing the ambient radiation's influence is presented in Dudzik (2007) and Dudzik and Minkina (2002). The investigated object was placed inside an open measurement chamber that reduced the influence of external radiation. In addition, the chamber walls were coated with an emulsion of high (close to unity) emissivity coefficient. Unfortunately, this method enables only a small class of objects to be investigated under laboratory conditions. Because, in general, the ambient temperature is unknown, in practical measurements it is assumed to be equal to the atmospheric temperature.

The influence of the atmosphere's self-radiation can be neglected when the camera-to-object distance does not exceed several meters. In temperature measurements of distant objects, the atmosphere's self-radiation should be taken into consideration. This is especially important when input variables ε_{ob}, T_o, T_{atm}, ω and d of the measurement model are correlated with each other (Dudzik 2005) and if, in addition, observed objects have low coefficients of emissivity (Dudzik and Minkina 2007).

The second source of errors arises in the calibration of infrared cameras (Hartmann *et al.* 2002). Temperature measurement error due to the calibration process usually arises from:

- differences in the self-radiation of the camera's optical components and filters during calibration and measurement and variations of this self-radiation with temperature;
- different camera-to-object distance during calibration and measurement;
- imprecise determination of the object emissivity during calibration, neglecting the influence of the ambient radiation reflected from the black body and limited temperature resolution of the camera;
- limited accuracy of the reference standard, and limited number of calibration points and interpolation errors.

Since the calibration process is an essential source of systematic interactions, we briefly describe its steps and theoretical foundations below.

Apart from automatic calibration of the detector, described in Chapter 3, the whole camera, as a final product, is sent by the manufacturer to a calibration laboratory where it undergoes a number of calibration steps. The calibration certificate is attached to each piece that passes this process. This certificate includes (among others):

- name of the laboratory;
- camera serial number;
- camera components (e.g. type of optics, filters) used during calibration;
- standard defining the calibration process – standard ITS-90 is currently in force;
- certificate validity period;
- date and signature of relevant persons (executing and approving person).

Technical data for each measurement camera specify its measurement accuracy, for example $\pm 2\,°C$ or $\pm 2\%$ of range. This parameter should be interpreted as accuracy under strictly specified laboratory conditions (e.g. during calibration). In practical measurements (e.g. outdoors) the accuracy may be significantly worse.

Calibration of a infrared system is conducted using a technical black body, whose $\varepsilon_{ob} \approx 1$ over the operating range of a calibrated device. For short distances between the camera and the black body we can assume that $TT_{atm} \approx 1$ and $\varepsilon_{ob} \approx 1$ (technical black body). Then, from (3.13) we get $s_{ob} = s$. Laboratory calibration is performed by measuring signals s_i corresponding to different temperatures T_i set on technical black bodies (Figure 2.2 and Figure 4.3), which are model sources of infrared radiation. The calibration curve, described by (4.3), is an approximation of points (s_i, T_i) by functions:

$$s_i = \frac{R}{\exp(B/T_i) - F}, \tag{4.3}$$

$$T_i = \frac{B}{\ln\left(\frac{R}{s_i} + F\right)}, \text{K}, \tag{4.4}$$

where R, B, F are constants determined to obtain the best fit of function (4.3) to the calibration points. On the basis of (4.4), the infrared camera converts detector signal s_i to temperature T_i for a specified range of radiation wavelengths (λ_1, λ_2).

From a theoretical point of view, characteristic (4.3) follows from an approximation of the integration of the product of the black body radiant exitance $M(\lambda,T)$ incident on the detector (according to Planck's law) and function $S_k(\lambda)$ describing the relative spectral sensitivity of the camera – Figure 4.1 (Wallin 1994, Kaplan 2000). This product is integrated over the camera operation interval λ_1–λ_2 and for specified temperature T_i of the black body:

$$s(T_i) = C \cdot \int_{\lambda_1}^{\lambda_2} S_k(\lambda) \cdot \frac{c_1 d\lambda}{\lambda^5 \cdot \left[\exp\left(\frac{c_2}{\lambda \cdot T_i}\right) - 1\right]}. \tag{4.5}$$

Characteristic $S_k(\lambda)$ is determined mainly by function $D^*(\lambda)$ of the detector's normalized detectivity (spectral sensitivity) and spectral characteristic of the camera optics transmission. In general, values of R, B, F are different for each camera piece and for each measurement range. Their values for the ThermaCAM PM 595 LW camera are presented in Table 4.1.

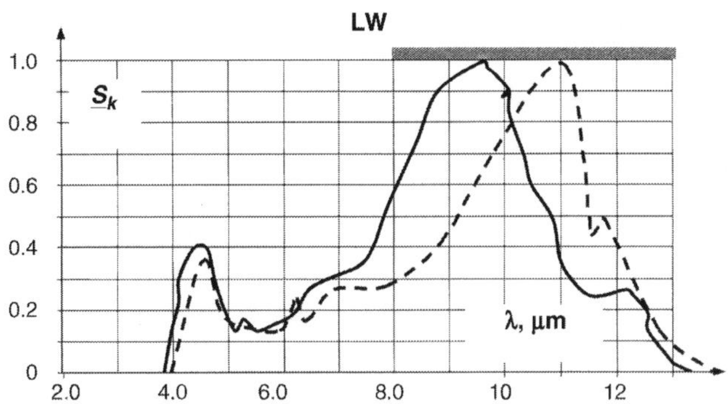

Figure 4.1 Example characteristics of relative spectral sensitivity $S_k(\lambda)$ of long-wave (LW) cameras (IR-Book 2000). Reproduced by permission of ITC Flir Systems

Table 4.1 Parameters *R*, *B*, *F* of calibration function of the ThermaCAM PM 595 (LW) camera (TOOLKIT IC2)

Measurement range, °C	*R*	*B*, K	*F*
−40–120	101 920	1463.4	1
80–500	17 250	1466.6	1
350–2000	1 870	1491.8	1

Typical static characteristics $s = f(T)$ of the measurement path of a short-wave (SW: 3–5 μm) and a long-wave (LW: 8–14 μm) camera are shown in Figure 4.2.

Constants *R*, *B*, *F* are stored in the memory of the camera's microcontroller, which calculates each time the measured temperature T_i on the basis of digitized detector signal *s*, measurement model (3.17) and set values of: object emissivity ε_{ob}, atmospheric temperature T_{atm}, ambient temperature T_o, camera-to-object distance *d* and atmospheric relative humidity $\omega_{\%}$.

The calibration procedure is described in detail in DeWitt (1983), Machin and Chu (2000) and Machin *et al.* (2008). The interior of a laboratory room for the calibration of infrared cameras and a set of technical black bodies are shown in the photos in Figure 4.3.

The third source of systematic errors is the electronic path of a camera. These errors result from the following unfavorable phenomena:

- detector noise;
- instability of the cooling system (in cameras with cooled arrays);
- fluctuations in gain of the preamplifier and/or other electronic systems;
- limited bandwidth of the detector and/or other electronic components;
- limited resolution and nonlinearity of A/D converters.

The errors in the electronic path are below ±1% for the ambient temperature from −15 to +40 °C.

In typical situations the errors in the method even reach several percent. They are the main source of errors in non-contact measurements of temperature fields with infrared cameras. Temperature measurement in infrared thermography is not accurate. It exhibits inaccuracies

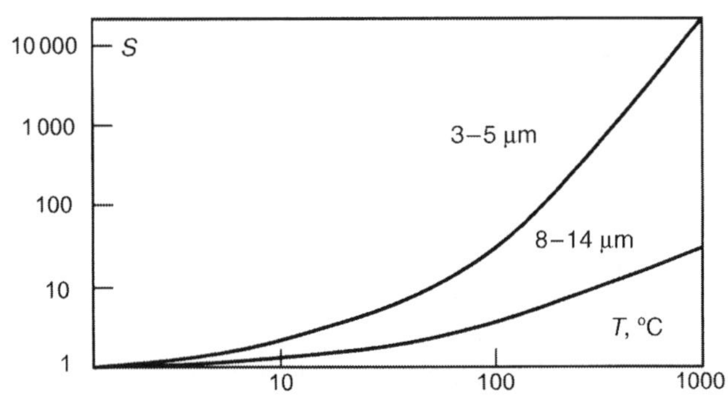

Figure 4.2 Examples of static characteristics of measurement path of a short-wave (SW: 3–5 μm) and long-wave (LW: 8–14 μm) camera

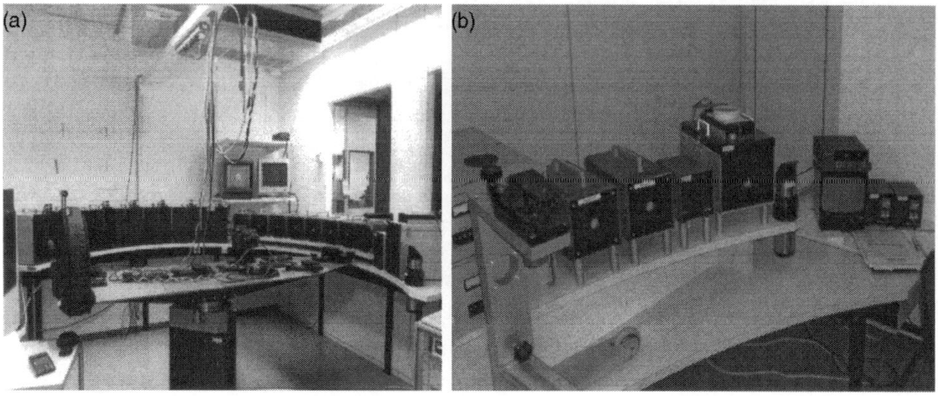

Figure 4.3 (a) Laboratory room for calibration of infrared cameras; (b) set of technical black bodies (IR-Book 2000). See Color Plate 7 for the color version. Reproduced by permission of ITC Flir Systems

similar to those of well-known and commonly used optical pyrometers. Contact methods with the use of thermoelectric, resistance or thermistor thermometers offer much better accuracy, especially as the range of measured temperature is similar for both methods. Unfortunately, contact methods sometimes cannot be applied. The inaccuracy of the infrared thermography method can be apparent, especially while measuring the temperature field of an inhomogeneous object built from materials of different emissivity. Therefore, this method is recommended for remote determination of the temperature distribution of homogeneous objects of very similar emissivity. Even though the resolution of a typical camera is about 0.05–0.1 K, the imaged temperature can be biased by many factors. Therefore, each result should be analyzed carefully and measurement staff should have vast experience in the interpretation of measurements in infrared thermography.

Since the error in the method is the main component in evaluating the accuracy of the measurement in infrared thermography, in a further part of this monograph we will extensively investigate such errors under different measurements conditions.

4.3 Simulations of Systematic Interactions

As mentioned earlier, knowledge of the measurement model is necessary for the evaluation of errors in the infrared thermography method. In this monograph we analyze the error in the method based on model (3.16), (3.17) with the atmospheric transmission defined by (3.15). Similar models are used by most manufacturers of infrared cameras. The difference is that each manufacturer uses its own model of TT_{atm}.

According to (3.17), five input values representing measurement conditions have to be defined in the camera software: object emissivity ε_{ob}, ambient temperature T_o, atmospheric temperature T_{atm}, relative humidity ω and camera-to-object distance d. Further in this section we analyze the influence of these input quantities on the bias of temperature measurement in infrared thermography. Because the measurement model (3.17) is highly nonlinear, we employ the (exact) method of increments, described in Chapter 1 (formula (1.8)), to evaluate the

Table 4.2 Reference values of the input parameters assumed in simulations of model (3.17)

Input parameter	Object emissivity (ε_{ob})	Ambient temperature (T_o), K	Atmospheric temperature (T_{atm}), K	Relative humidity (ω)	Camera-to-object distance (d)
Value	0.98; 0.80; 0.60; 0.40	293	293	0.5	1, 100

components of the bias associated with individual input parameters. The error in the method is investigated with respect to all the inputs of the measurement model. Simulations were performed in the MATLAB R2006b environment. At first, the measurement model was implemented in MATLAB scripts. Next, the created model was simulated for different measurement conditions (influencing inputs) and different input signals. The measurement conditions assumed in simulations are given in Table 4.2 and the ranges of relative error of the input parameters are listed in Table 4.3.

Work by Orlove (1982), DeWitt (1983) and Hamrelius (1991) deals with metrological analysis based on absolute errors. It is limited only to narrow intervals, because the generalization of results would condense graphs and make them unclear, thus making identification of changes in the error function impossible. In our opinion, using absolute values is less legible. For example, is absolute error $\Delta_{T_{ob}} = 50$ K big or small? If the object temperature $T_{ob} = 2000$ K, it seems to be small (relative error $\delta_{T_{ob}} = 2.5\%$) for a typical non-contact temperature measurement. Taking these inconveniences into consideration, we decided to conduct sensitivity analysis of the infrared camera measurement model using relative errors in the influencing quantities. We believe that simulation results, presented further as graphs of the relative error, are legible and make possible the drawing of interesting conclusions on the properties of the error function.

4.3.1 Influence of Emissivity Setting Error on Temperature Measurement Error

The problem of emissivity measurement and the impact of emissivity on the radiative properties of object surfaces was discussed in Section 2.3. In this section we present simulation results obtained for model (3.17) assuming that the object surface emissivity setting in the camera is incorrect.

The component of temperature measurement error $\delta_{T_{ob}}$ due to error $\delta_{\varepsilon_{ob}}$ of the object emissivity setting is presented in Figures 4.4 and 4.5. Graphs in Figure 4.4 refer to the case of $T_{ob} > T_o$ and $T_{ob} > T_{atm}$, while graphs in Figure 4.5 refer to the case of $T_{ob} \leq T_o$ and $T_{ob} \leq T_{atm}$.

For the case of $T_{ob} > T_o$ and $T_{ob} > T_{atm}$, simulations were performed for object temperatures equal to: 303 K (30 °C), 323 K (50 °C), 343 K (70 °C), 363 K (90 °C). For the case of $T_{ob} \leq T_o$

Table 4.3 Ranges of relative error of the input parameters of model (3.17)

Input parameter	Object emissivity (ε_{ob})	Ambient temperature (T_o), K	Atmospheric temperature (T_{atm}), K	Relative humidity (ω)	Camera-to-object distance (d)
Error range	±30%	±3%	±3%	±30%	±30

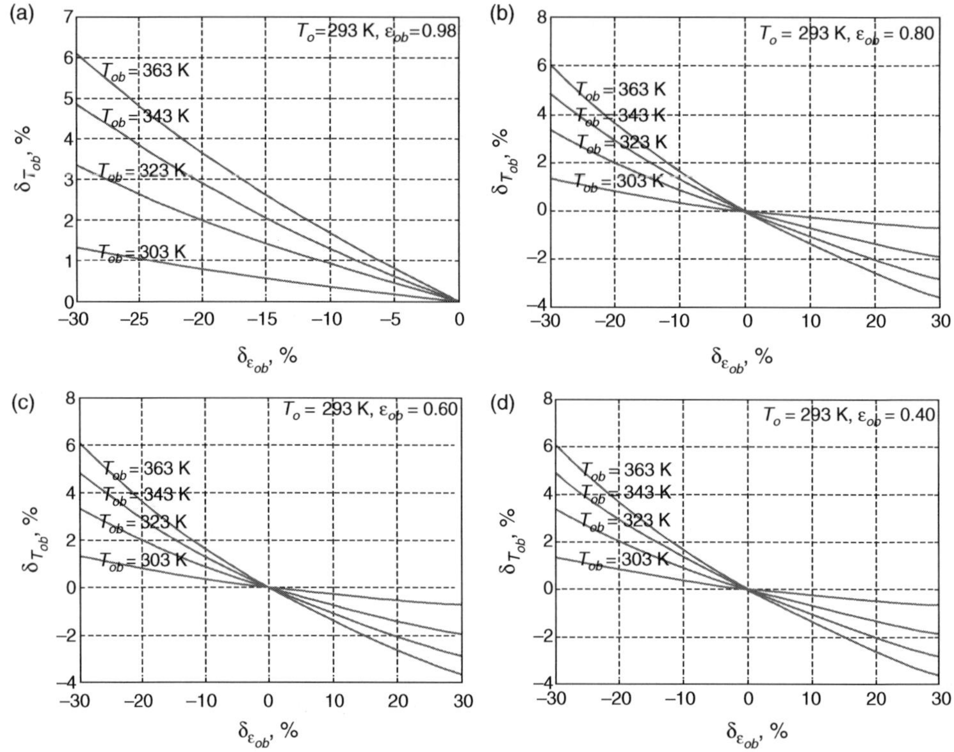

Figure 4.4 Influence of object emissivity ε_{ob} setting error on temperature T_{ob} measurement error ($T_{ob} > T_o$ and $T_{ob} > T_{atm}$) for $T_{atm} = T_o = 293$ K (Minkina and Bąbka 2002)

and $T_{ob} \leq T_{atm}$, simulations were performed for object temperatures equal to: 263 K (−10 °C), 274 K (1 °C), 283 K (10 °C), 293 K (20 °C).

On the basis of the presented results we can draw the following conclusions:

- The error in the object emissivity (ε_{ob}) setting has a strong impact on the error in temperature measurement. From Figure 4.4 we can see that overestimation of ε_{ob} produces a smaller error than underestimation. For example, for $\delta\varepsilon = +30\%$, error $\delta_{T_{ob}} \approx -(1\text{–}4)\%$, whereas for $\delta\varepsilon = -30\%$, error $\delta_{T_{ob}} \approx +(2\text{–}7)\%$ provided that: $T_{atm} = T_o = 293$ K, $T_{ob} = (300\text{–}400)$ K and $\varepsilon_{ob} = 0.4\text{–}0.98$. From Figure 4.4 we can also see that the considered error component increases with increasing T_{ob} and does not depend on ε_{ob}.
- From Figure 4.5 we can see that if $T_{ob} \leq T_o$ and $T_{ob} \leq T_{atm}$, then error $\delta_{T_{ob}} = f(\delta_{\varepsilon_{ob}})$ does not depend on ε_{ob} and increases strongly when T_{ob} decreases. Figure 4.5 illustrates the situation when the results of any infrared thermography measurements are unreliable. When $T_{ob} \leq T_o$ and $T_{ob} \leq T_{atm}$, measurements should not be carried out.
- Figures 4.4 and 4.5 show a change of sign in error $\delta_{T_{ob}}$ for $T_{ob} > T_o$, $T_{ob} > T_{atm}$ and for $T_{ob} \leq T_o$, $T_{ob} \leq T_{atm}$. When $T_{ob} \approx T_o$, $T_{ob} \approx T_{atm}$, error $\delta_{T_{ob}}$ is close to zero, which means that the model is less sensitive to changes in the considered quantity. For $T_{ob} = T_o = T_{atm}$ we have a critical case: the model is insensitive to changes of given input quantity *X*, where *X* stands

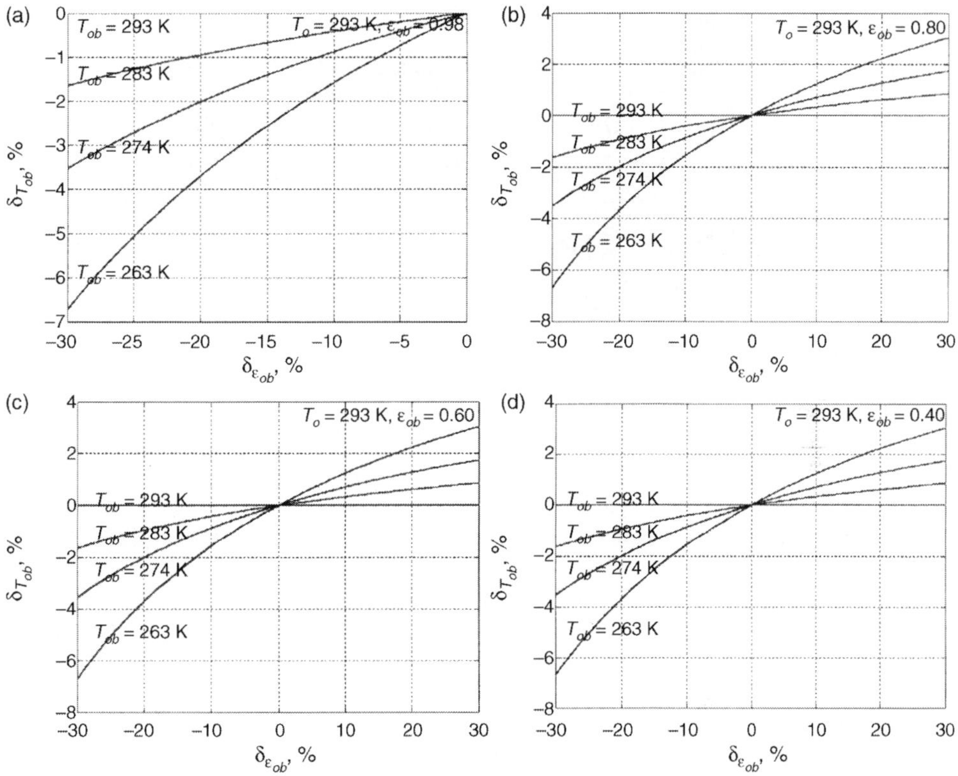

Figure 4.5 Influence of object emissivity ε_{ob} setting error on temperature T_{ob} measurement error ($T_{ob} \leq T_o$ and $T_{ob} \leq T_{atm}$) for $T_{atm} = T_o = 293$ K (Minkina 2004). Reproduced by permission of Częstochowa University of Technology

for ε_{ob}, T_o, T_{atm}, d or ω. We call it the singular point of the model. The described features of the model follow from the mathematical form of formula (3.17). On the other hand, when $T_{ob} = T_o = T_{atm}$, the error, according to the theory of measurement, tends to infinity. This confirms the above statement that the results of an infrared thermography measurement conducted when $T_{ob} = T_o = T_{atm}$ are unreliable.

In work by Minkina (2004), the model insensibility to changes of ε_{ob}, for $T_{ob} = T_o = T_{atm}$, is explained more vividly. Contributions from particular components of radiation arriving at the detector are presented graphically in Figure 4.6. The detector receives radiation fluxes from the object, atmosphere and ambient. Each of these fluxes is proportional to radiation intensity (which increases with temperature) and emissivity ε_{ob}, ε_{atm} and $\varepsilon_o = 1$, respectively. Figure 4.6a illustrates the situation when the object emissivity is small. In such a case, the measured intensity of radiation from the object makes up about 30% of the radiation received by the detector. The remaining 70%, coming from the atmosphere and ambient, comprises noise. Consequently, measurement conditions are very difficult. The situation shown in Figure 4.6b is even worse. The measurement conditions are extremely unfavorable not only due to the assumed temperature distribution, where the object is cooler than the ambient and the

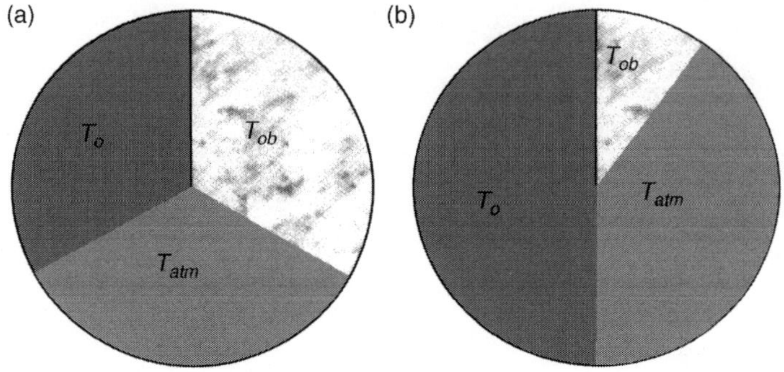

Figure 4.6 Graphical illustration of contributions from particular components of radiation arriving at the camera detector in the critical case

atmosphere ($T_{ob} \ll T_o$, $T_{ob} \ll T_{atm}$), but also due to the object emissivity, which is assumed smaller than the atmosphere and the ambient emissivity ($\varepsilon_{ob} \ll \varepsilon_{atm}$, $\varepsilon_{ob} \ll \varepsilon_o$). In such a case, the contribution of the object radiation to the total flux arriving at the camera detector can be even lower.

The simulation results presented above assume constant object temperature. Alternatively, we can investigate the influence of errors of individual input quantities in measurement model (3.17), assuming constant signal s_{ob} (Bąbka and Minkina 2002c), where:

$$s_{ob} = s\frac{1}{\varepsilon_{ob} TT_{atm}} - \left[\frac{1-\varepsilon_{ob}}{\varepsilon_{ob}}\frac{R}{\exp(B/T_o)-F} + \frac{1-TT_{atm}}{\varepsilon_{ob}\cdot TT_{atm}}\frac{R}{\exp(B/T_o)-F}\right]. \tag{4.6}$$

Graphs of the object temperature, calculated from the model, versus assumed object emissivity ε_{ob}, for constant s_{ob} and atmospheric transmission TT_{atm}, are presented in Figure 4.7.

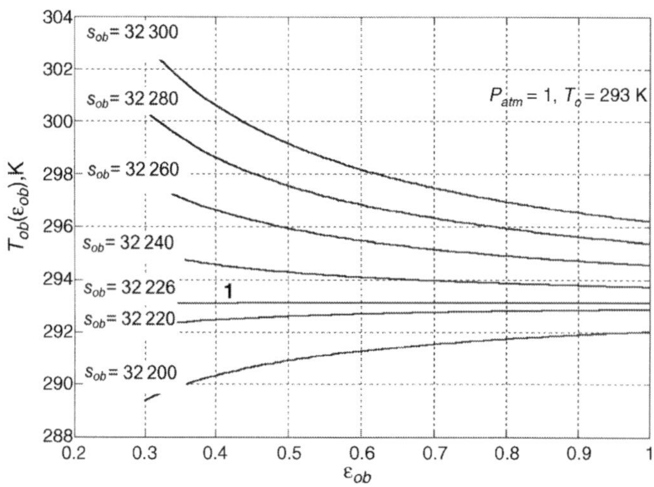

Figure 4.7 Influence of object emissivity ε_{ob} on object temperature measurement T_{ob} for constant s_{ob} (constant radiant exitance) (Bąbka and Minkina 2002c, Bąbka and Minkina 2003b)

The effect of the insensibility of T_{ob} to ε_{ob} for object temperature T_{ob} equal to ambient temperature T_o, described earlier, is represented by curve 1 in Figure 4.7. Other curves in Figure 4.7 indicate the dependence of T_{ob} on ε_{ob} for $T_{ob} \neq T_o$.

The case when the object temperature is close to the ambient temperature is not the only unfavorable situation occurring during measurements in infrared thermography. Another difficult situation is the temperature measurement of objects of low emissivity, like mirror surfaces of polished metals, when the interpretation of thermograms is much more difficult due to reflections.

The appearance of reflection is easy to detect by changing the angle of observation. After a small change in the angle, the radiation intensity of a hot object remains practically the same, whereas reflected radiation may change a lot. To illustrate problems in measurements in infrared thermography in the case of low emissivity of the investigated object surface, let us consider the following example.

Example 4.1 *Thermograms of object with low-emissivity surfaces*

The images in Figure 4.8 are thermograms of a polished aluminum sheet, whose temperature is equal to the ambient temperature. The emissivity of polished aluminum is $\varepsilon \approx 0.1$. Figure 4.8a shows the reflection of a person measuring the sheet temperature, while Figure 4.8b shows a glass

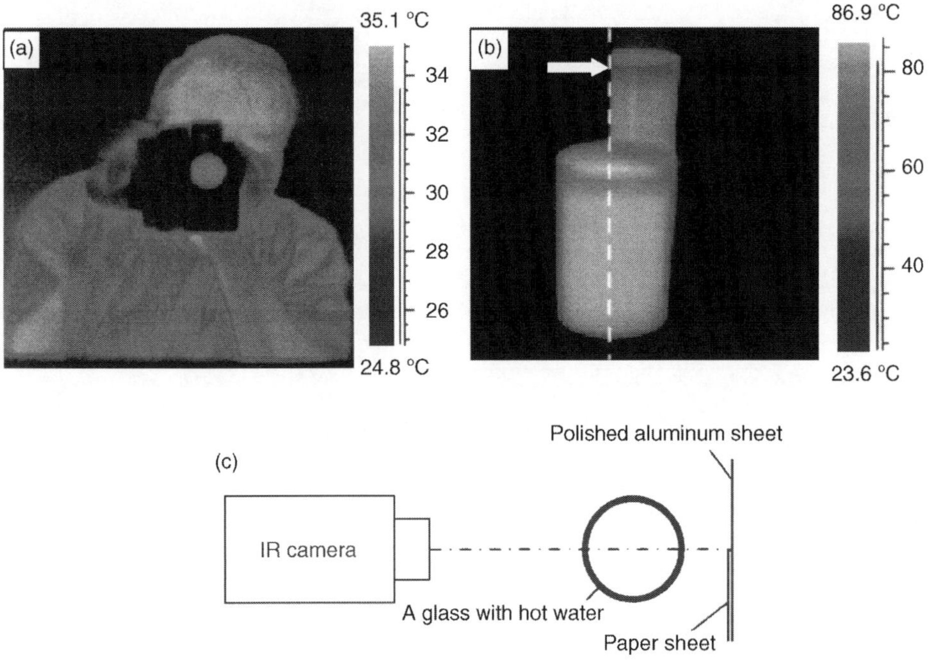

Figure 4.8 Thermograms of a polished aluminum sheet, whose temperature is close to the ambient temperature: (a) mirror reflection of a person measuring the sheet temperature; (b) image of a glass of hot water located against the sheet and its background reflection (right-hand edge of paper sheet stuck on the aluminum is marked with dashed line), (c) view (Minkina 2004). See Color Plate 8 for the color version. Reproduced by permission of Częstochowa University of Technology

with hot water located against the sheet and its background reflection. A sheet of white paper, of emissivity $\varepsilon \approx 0.8$, was stuck on the left-hand side of the aluminum for comparison. The right-hand edge of the paper is marked in the image with a white dashed line. As we can see, there is no thermal reflection in the glass from the paper sheet (the left part of the glass reflection is cut out along the paper edge). The apparent temperatures of the reflected objects (person, glass) are lower than the actual temperatures, because the aluminum sheet is not an ideal white body (its emissivity $\varepsilon_{ob} \approx 0.1$). If the aluminum emissivity were close to zero ($\varepsilon_{ob} \approx 0$), the recorded temperatures would be close to the actual ones. This example confirms the remarks made earlier: that infrared thermography measurements of low-emissivity objects, whose temperature is close to the ambient (atmosphere) temperature, are unreliable. Figure 4.8c shows the top view of the setup.

4.3.2 Influence of the Ambient Temperature Setting Error on Temperature Measurement Error

In this section we discuss the influence of an incorrect setting of ambient temperature T_o on temperature measurement error. The relevant simulations of measurement error $\delta_{T_{ob}}$ versus the ambient temperature error δ_{T_o} are shown in Figures 4.9 and 4.10. The results presented in

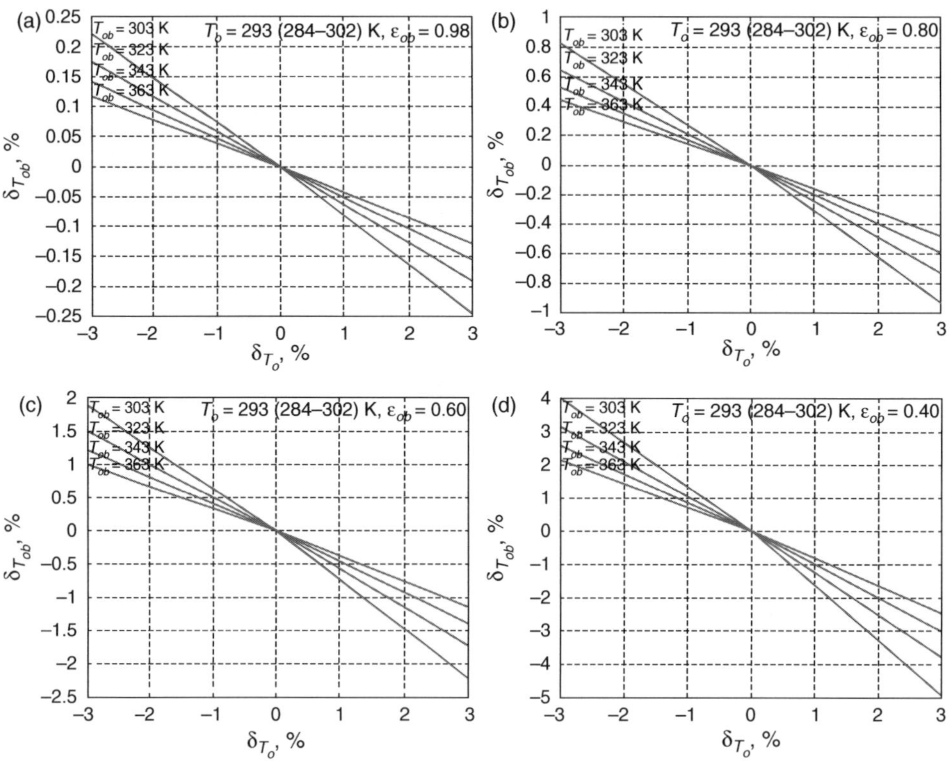

Figure 4.9 Influence of ambient temperature T_o setting error on temperature T_{ob} measurement error ($T_{ob} > T_o$ and $T_{ob} > T_{atm}$) for $T_{atm} = T_o = 293$ K (Minkina and Bąbka 2002, Minkina 2004). Reproduced by permission of Częstochowa University of Technology

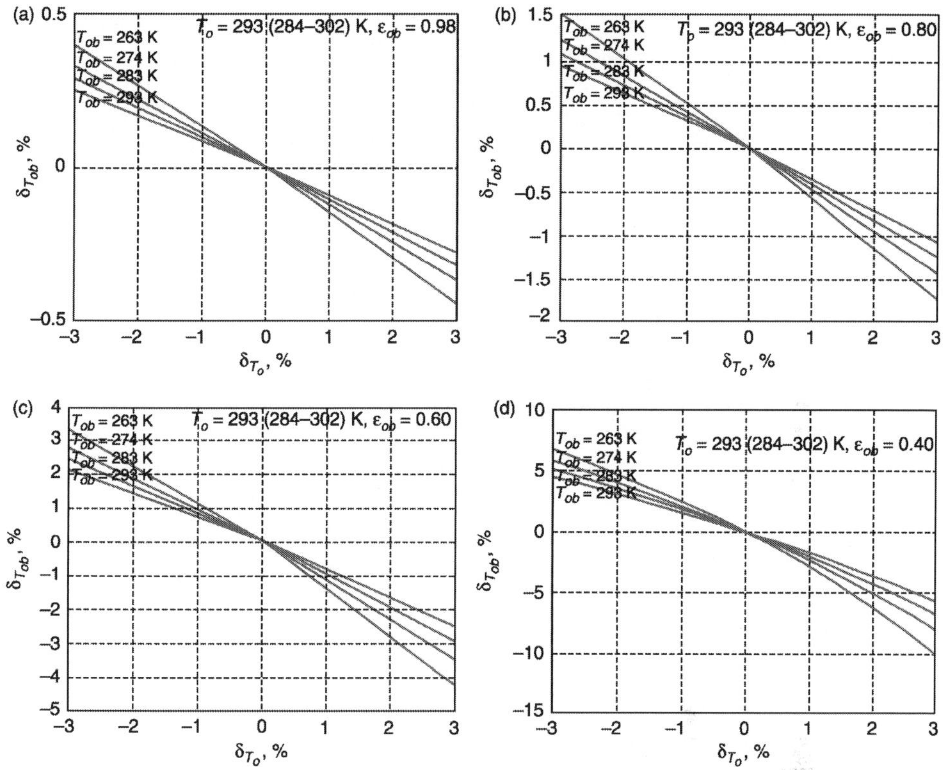

Figure 4.10 Influence of ambient temperature T_o setting error on temperature T_{ob} measurement error ($T_{ob} \leq T_o$ and $T_{ob} \leq T_{atm}$) for $T_{atm} = T_o = 293$ K (Minkina and Bąbka 2002, Minkina 2004). Reproduced by permission of Częstochowa University of Technology

Figure 4.9 were obtained for $T_{ob} > T_o$ and $T_{ob} > T_{atm}$; the results presented in Figure 4.10 for $T_{ob} \leq T_o$ and $T_{ob} \leq T_{atm}$.

Analysis of the above simulations of model (3.17), in terms of influence of incorrect setting of ambient temperature T_o on temperature measurement error, can be summed up as follows:

- Error δ_{T_o} in setting the ambient temperature also has a significant influence on the error of object temperature measurement T_{ob} (Figure 4.9), but not as strong as the incorrect setting of the object emissivity. As opposed to characteristics of δ_ε, characteristics of $\delta_{T_{ob}}$ versus δ_{T_o} exhibit symmetry, which means that both underestimated and overestimated settings of T_o result in similar errors. From Figures 4.9 and 4.10 we can see that overestimated setting leads to negative error and underestimated setting to positive error. For example, for $\delta_{T_o} = \pm 3\%$ error $\delta_{T_{ob}} \approx \pm(0.1\text{-}5.0)\%$, if $T_{atm} = T_o = 293$ K, $T_{ob} = (300\text{–}400)$ K and $\varepsilon_{ob} = 0.4\text{–}0.98$. We can also note that error δ_{T_o} decreases with increasing T_{ob} and ε_{ob}.
- Looking at Figure 4.10, we can see that, if $T_{ob} \leq T_o$ and $T_{ob} \leq T_{atm}$, error $\delta_{T_{ob}} = f(\delta_{T_o})$ increases strongly when ε_{ob} and T_{ob} decrease. This is the situation described earlier and illustrated with graphs of the error component associated with incorrect setting of the object

emissivity (Figure 4.5). In such situations the results of infrared thermography measurements cannot be regarded as reliable.

- Figure 4.9a shows that for high measured temperatures the influence of the ambient temperature on measurement results can be neglected, especially if the object emissivity is high. This is an important conclusion for practical measurements in infrared thermography.

4.3.3 *Influence of the Atmospheric Temperature Setting Error on Temperature Measurement Error*

The influence of an incorrect setting of atmospheric temperature T_{atm} on temperature measurement error is presented in Figures 4.11 and 4.12. The graphs show simulation results of measurement error $\delta_{T_{ob}}$ versus the atmospheric temperature error $\delta_{T_{atm}}$, carried out for measurement model (3.17) and the atmospheric transmission (TT_{atm}) model (3.15).

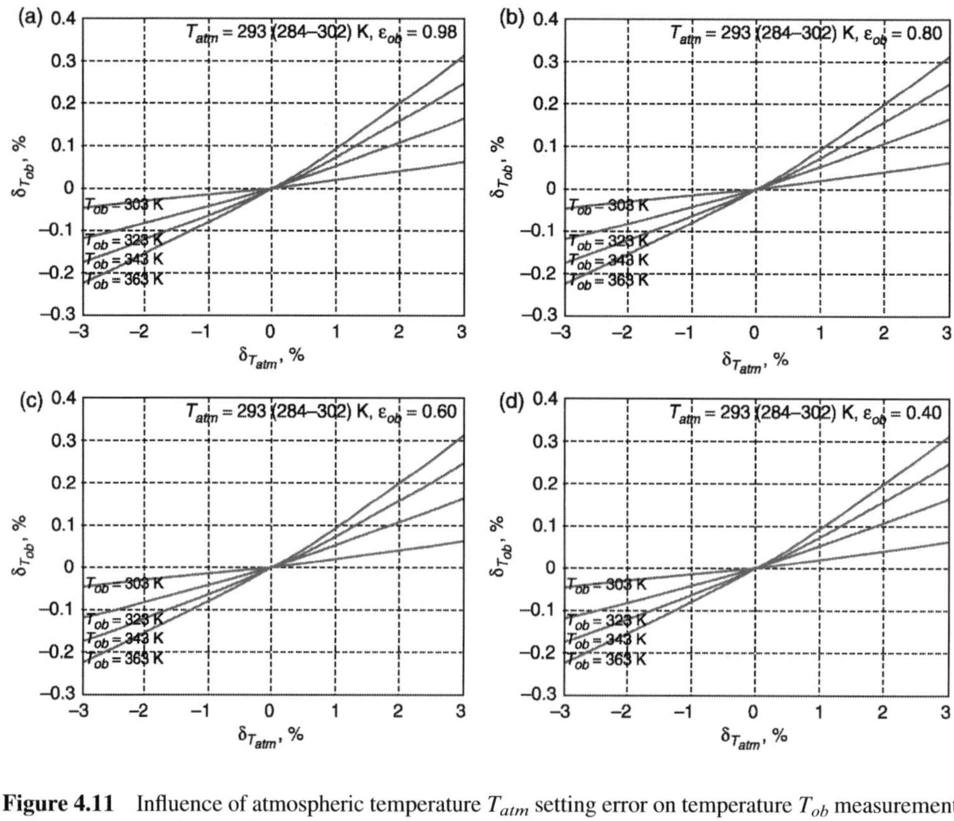

Figure 4.11 Influence of atmospheric temperature T_{atm} setting error on temperature T_{ob} measurement error ($T_{ob} > T_o$ and $T_{ob} > T_{atm}$) for $T_{atm} = T_o = 293$ K (Minkina 2004). Reproduced by permission of Częstochowa University of Technology

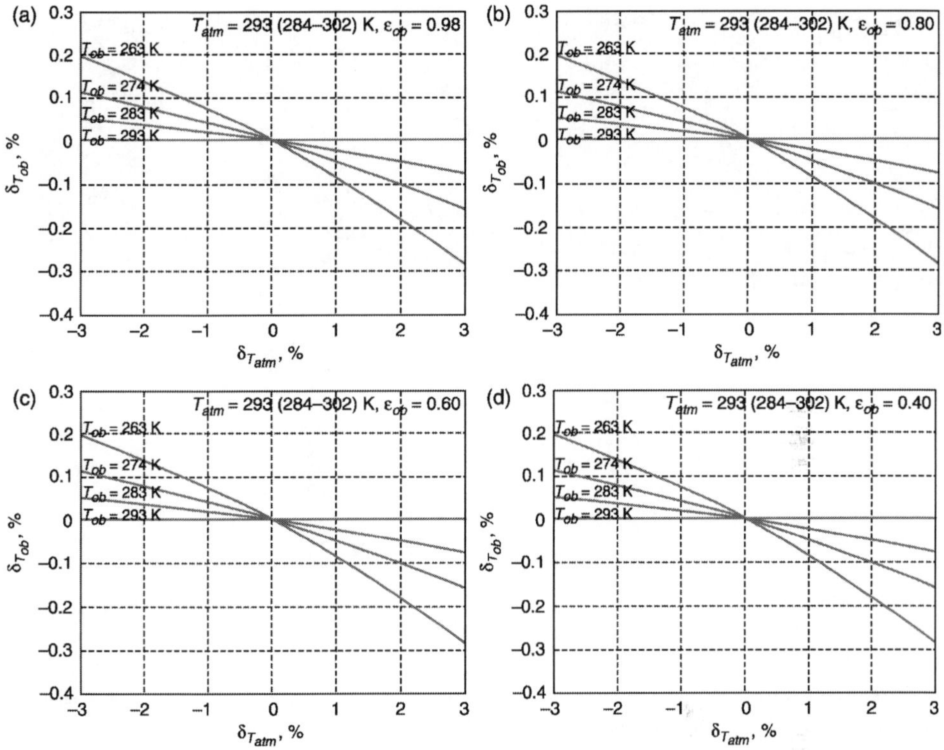

Figure 4.12 Influence of atmospheric temperature T_{atm} setting error on temperature T_{ob} measurement error ($T_{ob} \leq T_o$ and $T_{ob} \leq T_{atm}$) for $T_{atm} = T_o = 293$ K (Minkina 2004). Reproduced by permission of Częstochowa University of Technology

Analyzing the results shown in Figures 4.11 and 4.12, in terms of influence of incorrect setting of atmospheric temperature on temperature measurement error, we can state that:

- Incorrect setting of atmospheric temperature T_{atm} does not greatly affect the error of the object temperature measurement – Figure 4.11. The error in T_{ob} measurement is proportional to the error in T_{atm} setting. For example, for $\delta_{T_{atm}} = \pm 3\%$ error, $\delta_{T_{ob}} \approx \pm(0.05\text{–}0.35)\%$ if $T_o = T_{atm} = 293$ K, $T_{ob} = (300\text{–}400)$ K and $\varepsilon_{ob} = 0.4\text{–}0.98$. The error increases with T_{ob} and does not depend on ε_{ob}.
- If $T_{ob} \leq T_{atm}$ and $T_{ob} \leq T_o$, we can see from Figure 4.12 that error $\delta_{T_{ob}} = f(\delta_{T_{atm}})$ is independent of ε_{ob} and increases when T_{ob} decreases. Even though this component of the temperature measurement error is not large, measurements ought not to be performed under such conditions, as follows from the general principles of infrared thermography measurements described earlier.

Example 4.2 *Influence of incorrect settings of the ambient and the atmospheric temperature on temperature measurement of an industrial installation*

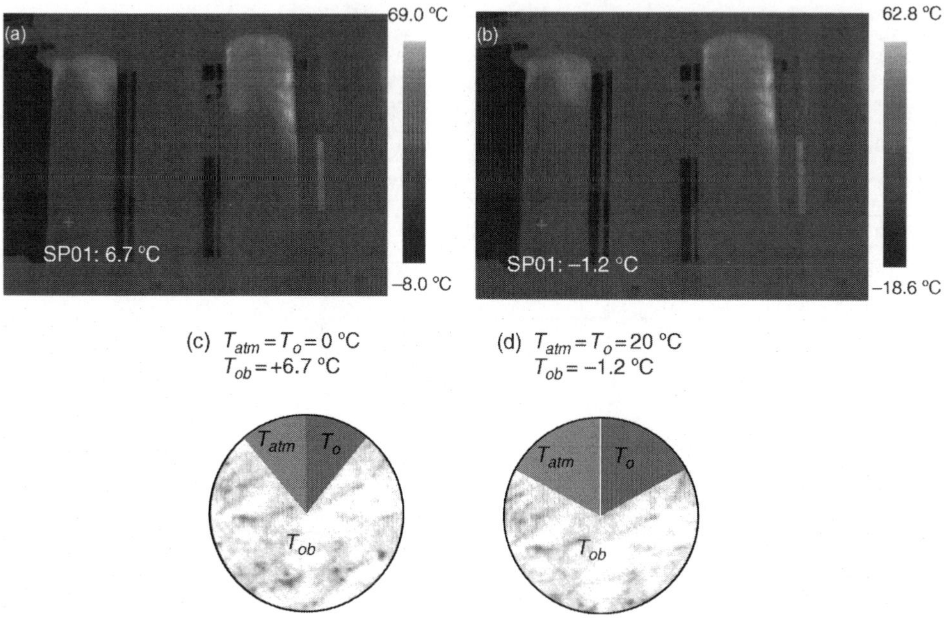

Figure 4.13 Influence of incorrect settings of T_{atm} and T_o on temperature measurement of an industrial installation (Minkina 2003). See Color Plate 9 for the color version. Reproduced by permission of Częstochowa University of Technology

The result of incorrect settings of the ambient and atmospheric temperature on the interpretation of an industrial installation thermogram is presented in Figures 4.13a and b. The object under investigation was the lining of an industrial furnace chimney. To illustrate the error following incorrect settings of T_{atm} and T_o in the camera, we first entered correct values $T_{atm} = T_o = 0\,°C$ and the recorded thermogram is shown in Figure 4.13a. The camera calculated the object temperature (the lining temperature) at the marked spot as $T_{ob} = +6.7\,°C$. Next, we changed the settings to incorrect values $T_{atm} = T_o = +20\,°C$ and the recorded thermogram is shown in Figure 4.13b. This time the camera calculated the object temperature at the same spot as $T_{ob} = -1.2\,°C$. In both cases the camera software calculated T_{ob} for the same emissivity of the lining, $\varepsilon = 0.8$. The difference in indications arises from the fact that the camera measures the total radiation intensity arriving at each pixel of the detector. We can analyze the contributions of particular components of radiation arriving at the detector for incorrect settings of the ambient and the atmospheric temperature (Figures 4.13c and d), as was done earlier for an incorrect setting of the object emissivity. When the settings are correct, i.e. $T_{atm} = T_o = 0\,°C$, the camera's interpretation of the radiation components is correct: the detector receives more radiation from the object and less from the ambient and the atmosphere (Figure 4.13c), so the calculated object temperature is higher ($T_{ob} = +6.7\,°C$). When the settings are incorrect, $T_{atm} = T_o = 20\,°C$ (i.e. the temperatures are set too high), the camera underrates the contribution of the object radiation and overrates contributions from the ambient and the atmosphere (Figure 4.13d). Consequently, the camera indication is too low ($T_{ob} = -1.2\,°C$).

4.3.4 *Influence of Camera-to-object Distance Setting Error on Temperature Measurement Error*

In this monograph we investigate the influence of atmospheric transmission on measurement error in infrared thermography considering the influence of the camera-to-object distance and the atmospheric relative humidity. As described in Section 3.1, the atmosphere has limited transmission that depends on infrared radiation wavelength λ, atmospheric temperature T_{atm}, humidity ω and camera-to-object distance d (Figures 3.1a and b). In this section we deal with the dependence of temperature measurement error on incorrect setting of the camera-to-object distance. The influence of incorrect setting of the relative humidity is discussed in the next section.

Simulation results in Figures 4.14 and 4.15, carried out for measurement model (3.17) and the atmospheric transmission (TT_{atm}) model (3.15), show the relationship of temperature measurement error $\delta_{T_{ob}}$ to camera-to-object distance error δ_d. The graphs in Figure 4.14 are for $T_{ob} > T_o$ and $T_{ob} > T_{atm}$ and the graphs in Figure 4.15 for $T_{ob} \leq T_o$ and $T_{ob} \leq T_{atm}$.

Analysis of the presented results in terms of influence of the error component associated with camera-to-object distance allows us to draw the following conclusions:

- The camera-to-object distance error has minimal influence on the measurement error of the object temperature: $\delta_{T_{ob}} < \pm 0.2\%$ for $\delta_d = \pm 30\%$ if $T_o = T_{atm} = 293$ K, $T_{ob} = (300–400)$ K

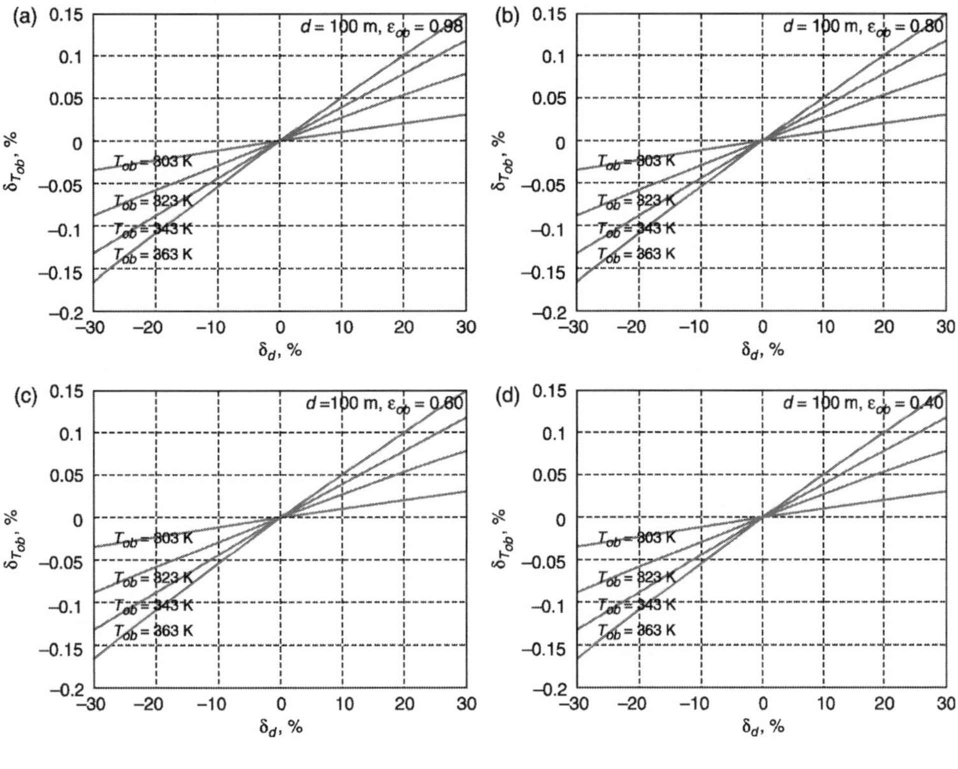

Figure 4.14 Influence of camera-to-object distance d setting error on temperature T_{ob} measurement error ($T_{ob} > T_o$ and $T_{ob} > T_{atm}$) for $T_{atm} = T_o = 293$ K (Minkina and Bąbka 2002)

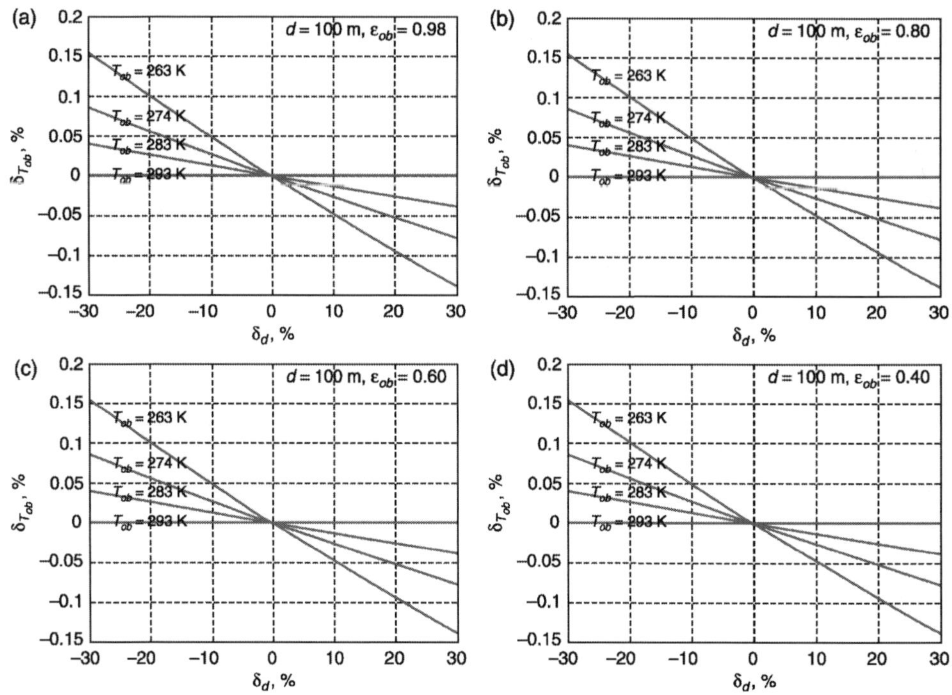

Figure 4.15 Influence of camera-to-object distance d setting error on temperature T_{ob} measurement error ($T_{ob} \le T_o$ and $T_{ob} \le T_{atm}$) for $T_{atm} = T_o = 293$ K (Minkina 2004). Reproduced by permission of Częstochowa University of Technology

and $\varepsilon_{ob} = 0.4$–0.98 (Figure 4.14). Looking at Figure 4.14, we can see that this component of the error increases with object temperature T_{ob} and does not depend on emissivity ε_{ob}.

- Characteristics in Figure 4.15 show that for $T_{ob} \le T_{atm}$ and $T_{ob} \le T_o$, error $\delta_{T_{ob}} = f(\delta_d)$ is also independent of ε_{ob} and increases when T_{ob} decreases. Even though temperature errors are small, measurements ought not to be carried out under these conditions, as follows from the general principles of measurements in infrared thermography described earlier.

4.3.5 Influence of Relative Humidity Setting Error on Temperature Measurement Error

Simulation results concerning the relationship of temperature measurement error $\delta_{T_{ob}}$ to relative humidity error δ_ω are presented in Figures 4.16 and 4.17. The results in Figure 4.16 are for $T_{ob} > T_o$ and $T_{ob} > T_{atm}$ and the results in Figure 4.17 for $T_{ob} \le T_o$ and $T_{ob} \le T_{atm}$.

Analysis of Figures 4.16 and 4.17 allows us to draw the following conclusions:

- Figure 4.16 shows that relative humidity error has minimal influence on the object temperature measurement error: $\delta_{T_{ob}} < \pm 0.2\%$ for $\delta_\omega = \pm 30\%$ if $T_o = T_{atm} = 293$ K, $T_{ob} =$ (300–400) K and $\varepsilon_{ob} = 0.4$–0.98. This component of the error increases with object

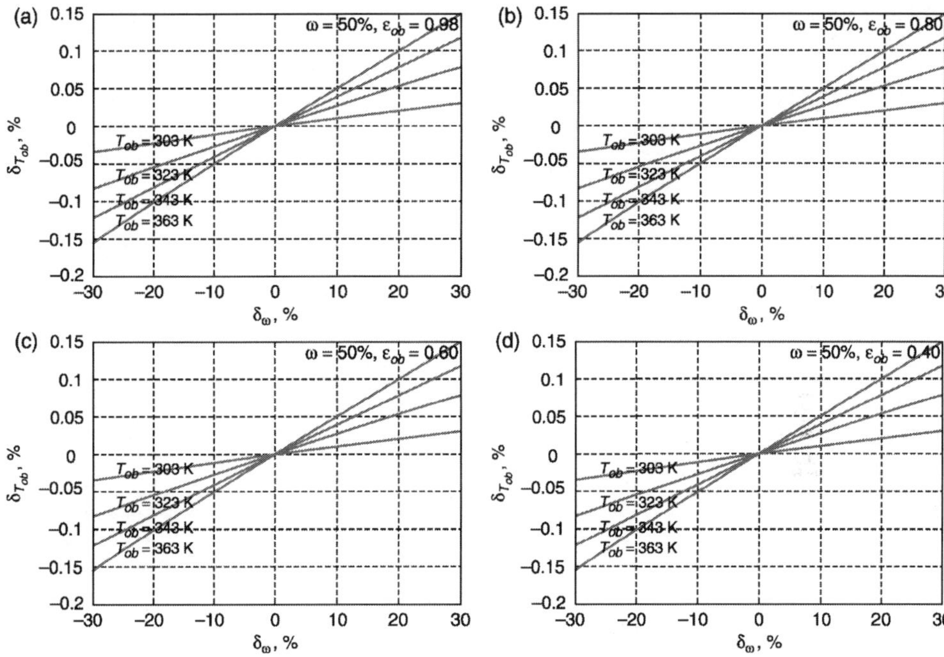

Figure 4.16 Influence of the atmospheric relative humidity ω setting error on temperature T_{ob} measurement error ($T_{ob} > T_o$ and $T_{ob} > T_{atm}$) for $T_{atm} = T_o = 293$ K (Minkina 2004). Reproduced by permission of Częstochowa University of Technology

temperature T_{ob} and does not depend on emissivity ε_{ob}. Its characteristics are very similar to that of the camera-to-object distance d error component (Figure 4.14).

- From Figure 4.17 we can see that for $T_{ob} \leq T_{atm}$ and $T_{ob} \leq T_o$, error $\delta_{T_{ob}} = f(\delta_\omega)$ is independent of emissivity ε_{ob} and increases when T_{ob} decreases. The graphs in Figure 4.17 are very similar to the graphs in Figure 4.15. Again, even though temperature errors are small, measurements ought not to be carried out under such conditions.

4.3.6 Summary

Summing up the chapter concerning the simulation investigation of errors in infrared thermography due to systematic interactions, we can state that the main component of temperature measurement error results from incorrect setting of the object emissivity. Therefore, in the case of objects of low emissivity, infrared thermography measurement is in principle unreliable or even impossible. Another critical case is the situation where the temperature of an object is similar to the ambient and atmospheric temperature. In such a case, radiation of the background (noise) may be much stronger than radiation of the investigated object (useful signal). In practice, the object temperature should be greater by at least 50 °C than the background temperature. In our practical experience, those situations where the temperature of a measured object is much higher than the temperature of the atmosphere or ambient occur most frequently. Evaluation of the unknown temperature of the ambient is a quite different problem. In practice, it is assumed that $T_o = T_{atm}$.

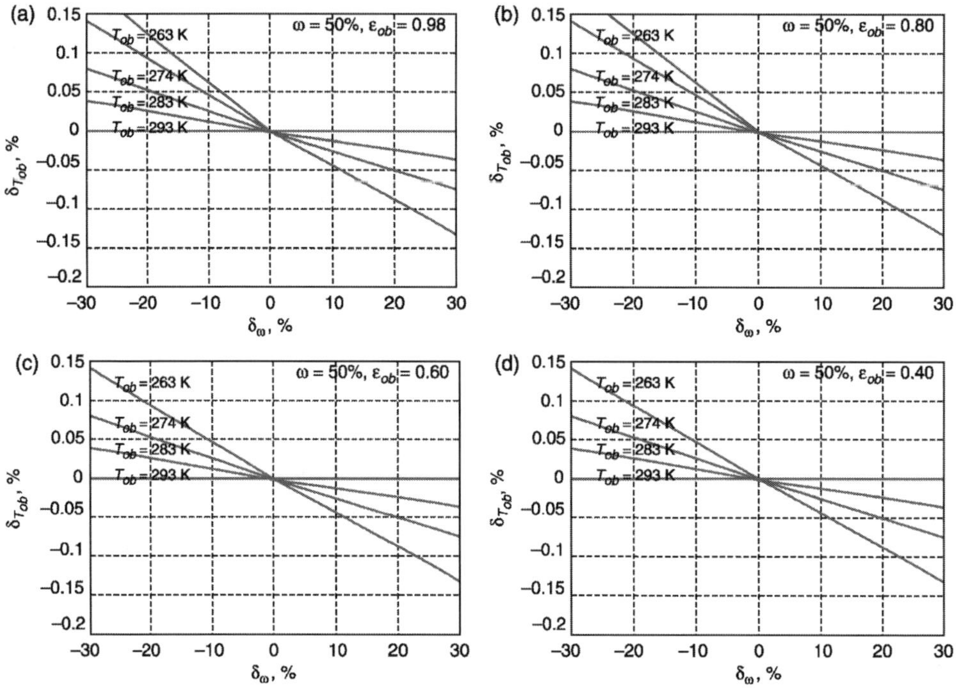

Figure 4.17 Influence of the atmospheric relative humidity ω setting error on temperature T_{ob} measurement error ($T_{ob} \leq T_o$ and $T_{ob} \leq T_{atm}$) for $T_{atm} = T_o = 293$ K (Minkina 2004). Reproduced by permission of Częstochowa University of Technology

All the above remarks lead to the conclusion that evaluation of errors in infrared thermography measurement should be treated on an individual basis, depending on the particular case. Theoretical analysis of such errors is still an open issue.

The resulting $\delta_{T_{ob}}$, obtained in this chapter, concerns the ThermaCAM PM 595 LW, FLIR camera. However, model (3.17) is valid for other cameras. Therefore, $\delta_{T_{ob}}$ and the conclusions will be similar for most cameras produced in the world.

As mentioned earlier, the theory of errors is not the only method of quantitative analysis of measurement inaccuracy. In this monograph we present another approach, based on the concept of uncertainty. In this approach all the input quantities of model (3.17) and the atmospheric transmission TT_{atm} model (3.15) are treated as random variables. The methodology and results of the uncertainty investigation of measurement in an infrared thermography model are presented in Chapter 5.

5

Uncertainties of Measurements in Infrared Thermography

5.1 Introduction

In Chapter 1 we presented the basics of the theory of measurement errors and uncertainties with particular attention to indirect measurements. In such cases, a measurement model is represented by a function, usually, of many variables. In Chapter 4 we used the model of infrared thermography measurement defined by a function of five variables (3.17) and the method of increments (described in Chapter 1) to evaluate the influence of errors in individual input variables on the output variable error. A model of accuracy defined in this manner is fully deterministic. The approach presented in Chapter 4 allows for analysis of a measurement model from the point of view of its sensitivity to given (determined) deviations of input quantities. In fact, sources of inaccuracy should be searched for not only in features of a model itself, but in the structure of measurement data (representing input quantities) as well. In the theory of measurement this is described as random interactions. In a deterministic model of accuracy (with systematic interactions), all the information on error is represented by particular values of individual input quantities. This does not encompass all possible interdependencies among inputs so, consequently, analysis deals each time with one particular set of input variables.

Another approach to the problem of evaluating accuracy is based on the modern theory of uncertainties (Guide 1995). Basic definitions of this theory, as well as the method of uncertainty evaluation, are presented in Chapter 1. As follows from that chapter, uncertainty is a statistical measure. A model of accuracy based on the concept of uncertainty can be called a random model. A basic feature that differentiates a statistical model from a deterministic one is the possibility of taking into consideration at the same time the structure of input data, represented by a probability density function or a cumulative distribution function, and the properties of a measurement model. When we use the theory of uncertainties, the evaluation of accuracy is based not on a single vector of input data, but on taking into account information included in large sets of representative input data. Therefore, it is necessary to define these sets as random variables. Unfortunately, statistical parameters determining accuracy on the basis of the theory of uncertainty are random variables as well. Consequently, they are evaluated only with some probability.

In this work, the investigation of temperature measurement uncertainty is based on the idea of **data processing algorithm uncertainty** (Minkina and Dudzik 2005), which is a measure of spread of an output random variable, equal to the standard experimental deviation of this variablc.

5.2 Methodology of Simulation Experiments

For a proper evaluation of measurement uncertainty, we make the following assumptions (Minkina and Dudzik 2005, Minkina and Dudzik 2006b):

- The input quantities of model (3.17) are random variables of given frequency distribution; they will be further called the input variables.
- The uncertainty of the object temperature measurement described by model (3.17) is a measure of spread of the output quantity realized around the expected value of this quantity.
- Expected values and the measure of spread of the input variables are modeled by the distribution parameters of these variables: the arithmetic mean and the standard experimental deviation.
- For a big enough number of input variable realizations, the arithmetic mean and the standard experimental deviation are unbiased estimators of the expected value and the standard deviation respectively (Guide 1995, Skubis 2003).
- Simulations were carried out for two options: the model input variables are uncorrelated; the model input variables are mutually correlated.

In simulations we used the Monte Carlo method, in accordance with the recommendations of Working Group No. 1 of the Joint Committee for Guides in Metrology (JCGM) (Guide 2004). The methodology of research leading to the evaluation of combined standard uncertainty components comprised the following steps:

1. Evaluation of the distribution parameters of the input variables.
2. Generation of series of input variable realizations for parameters evaluated in step 1 and for a defined level of correlation of the variables.
3. Simulation of the measurement model for data series generated in step 2.
4. Analysis of simulation results.

Besides the determination of uncertainty components, we also performed simulations intended to evaluate the combined standard uncertainty and the 95% coverage interval, according to the procedure described in Section 1.3.

In the analysis we assumed one of the two distributions of the input variables: the logarithmic Gaussian or the uniform distribution. The uniform distribution was applied to investigate critical cases, whereas the logarithmic Gaussian distribution was used to ensure stability of the numerical algorithm for generated random variables. To ensure the correct simulation of model (3.17), values of the input quantities cannot be negative.

5.2.1 Evaluation of Distribution Parameters of Input Variables

5.2.1.1 Logarithmic Gaussian Distribution

The probability density function of the logarithmic Gaussian distribution is given as:

$$p(z) = \frac{1}{z \cdot s \cdot \sqrt{2 \cdot \pi}} \exp^{\left(\frac{-(\ln(z)-m)^2}{2 \cdot s^2}\right)}, \tag{5.1}$$

where m and s are probability distribution parameters.

The expected value of random variable Z is defined as (Guide 1995):

$$E(Z) = \int z \cdot p(z) dz. \tag{5.2}$$

On the basis of (5.1) and (5.2), the expected value of variable Z is subject to the logarithmic Gaussian distribution:

$$E(Z) = \exp\left(m + \frac{s^2}{2}\right). \tag{5.3}$$

The variance of random variable Z is defined as:

$$V(Z) = E\{[Z - E(Z)]^2\}. \tag{5.4}$$

From formulas (5.1) and (5.4), the variance of variable Z is subject to the logarithmic Gaussian distribution:

$$V(Z) = \exp(2 \cdot m + 2 \cdot s^2) - \exp(2 \cdot m + s^2). \tag{5.5}$$

Solving the system of equations (5.3) and (5.5) for parameters m and s yields:

$$\begin{cases} m = \ln\left(\dfrac{E^2(Z)}{\sqrt{V(Z) + E^2(Z)}}\right) \\ s = \ln\left(\sqrt{\dfrac{V(Z) + E^2(Z)}{E^2(Z)}}\right) \end{cases}. \tag{5.6}$$

Using equations (5.6) we can determine such values of parameters m and s of the logarithmic Gaussian distribution that the expected value and the variance of random variable Z are equal to $E(Z)$ and $V(Z)$ respectively.

5.2.1.2 Uniform Distribution

The probability density function of the uniform distribution is given as:

$$p(z) = \begin{cases} \dfrac{1}{b-a} & \text{for } a \leq z \leq b \\ 0 & \text{for other } z, \end{cases} \tag{5.7}$$

Table 5.1 Equations defining parameters of distributions used in simulation analysis of model (3.17) sensitivity

Distribution	Expected value $E(z)$	Variance $V(z)$	Distribution parameters
Logarithmic Gaussian	$\exp\left(m+\frac{s^2}{2}\right)$	$\exp(2\cdot m+2\cdot s^2) - \exp(2\cdot m+s^2)$	$\begin{cases} m=\ln\left(\frac{E^2(Z)}{\sqrt{V(Z)+E^2(Z)}}\right) \\ s=\ln\left(\sqrt{\frac{V(Z)+E^2(Z)}{E^2(Z)}}\right) \end{cases}$
Uniform	$E(Z)=\frac{a+b}{2}$	$V(Z)=\frac{(b-a)^2}{12}$	$\begin{cases} a=E(Z)-\sqrt{3\cdot V(Z)} \\ b=E(Z)+\sqrt{3\cdot V(Z)} \end{cases}$

where a and b are distribution parameters. On the basis of (5.2), the expected value of the uniform distribution is:

$$E(Z)=\frac{a+b}{2} \tag{5.8}$$

and the variance is:

$$V(Z)=\frac{(b-a)^2}{12}. \tag{5.9}$$

Solving equations (5.8) and (5.9) for a and b, we can determine, as for the logarithmic Gaussian distribution, relationships between parameters a, b and the expected value and variance of the uniform distribution:

$$\begin{cases} a=E(Z)-\sqrt{3\cdot V(Z)} \\ b=E(Z)+\sqrt{3\cdot V(Z)}. \end{cases} \tag{5.10}$$

Equations describing the dependence of parameters of the two used distributions on their statistics are presented in Table 5.1 (Minkina and Dudzik 2006a).

5.2.2 Generation of Series of Input Variable Realizations

As mentioned above, the random input variables of the measurement model were generated for two variants of simulation. The first one assumes no correlation between the variables, the second one generates realizations for specified levels of correlation among the variables.

5.2.3 Uncorrelated Input Variables

Realizations of the input random variables, representing individual input quantities of measurement model (3.17) and the atmospheric transmission model (3.15), were generated using built-in pseudorandom generator functions of the MATLAB environment. They allowed for the generation of series of realizations that are subject to specified probability density

Table 5.2 Estimates of the inputs of model (3.17) assumed for simulations of components of the combined standard uncertainty

Input quantity	Object emissivity (ε_{ob})	Ambient temperature (T_o), K	Atmospheric temperature (T_{atm}), K	Relative humidity (ω)	Camera-to-object distance (d), m
Estimate value	0.9, 0.8, 0.6, 0.4	293	293	0.5	1, 100

functions. Parameters of the generators are simultaneously parameters of the probability density functions, calculated earlier from (5.6) – for the logarithmic Gaussian distribution – or from (5.10) – for the uniform distribution. The input, right-hand side data for (5.6) and (5.10) were specified a priori statistics of a given input variable, namely its standard uncertainty and expected value.

Values of estimates of the five input variables are given in Table 5.2. To evaluate the influence of standard uncertainties of the inputs on the uncertainty of the object temperature evaluation for model (3.17) with model (3.15), we have to define ranges of change of these uncertainties. The ranges of change of the input variables relative to standard uncertainties, used for simulations presented in this work, are given in Table 5.3. The percentage ranges of the ambient and atmospheric temperature are specified for the absolute Kelvin scale, according to the recommendation of the International Temperature Scale (ITS-90).

In the first step of the research, we assumed a lack of correlation among the input variables and analyzed the worst possible case, that is the uniform distributions of the variables. Series of $N = 10\,000$ input realizations were generated for the assumed expected values and experimental standard deviations. Figures 5.1–5.5 show histograms corresponding to probability density distributions $g(x_i)$ (where x_i is the ith input variable) of the generated inputs. The histograms are 20-bin normalized histograms (the sum of all heights of the bins is equal to one). The realizations were generated using MATLAB's uniform random generator. Calibration parameters and measurement conditions used in the simulations were read from thermogram files, recorded during experiments with the FLIR ThermaCAM PM 595 LW infrared camera. The calibration parameters relate to the measurement range −40 to 120 °C. The symbols in the graphs are E for expected value and σ for standard deviation.

Figures 5.6–5.10 show histograms of the input variable realizations subject to the logarithmic Gaussian distribution. They were generated for the same statistics (expected value and standard deviation) as before and used for simulating components of the joint uncertainty.

Table 5.3 Ranges of relative standard uncertainties of the model (3.17) assumed for simulations of components of the combined standard uncertainty

Input quantity	Object emissivity (ε_{ob})	Ambient temperature (T_o), K	Atmospheric temperature (T_{atm}), K	Relative humidity (ω)	Camera-to-object distance (d), m
Uncertainty range (%)	0–30	0–3	0–3	0–30	0–30

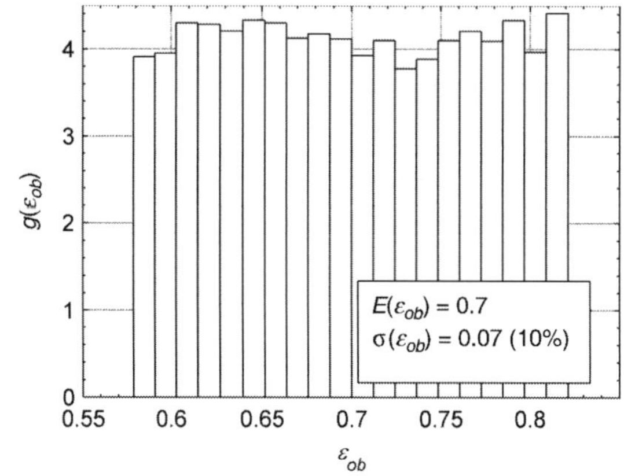

Figure 5.1 Probability density function of variable representing emissivity ε_{ob} (uniform distribution)

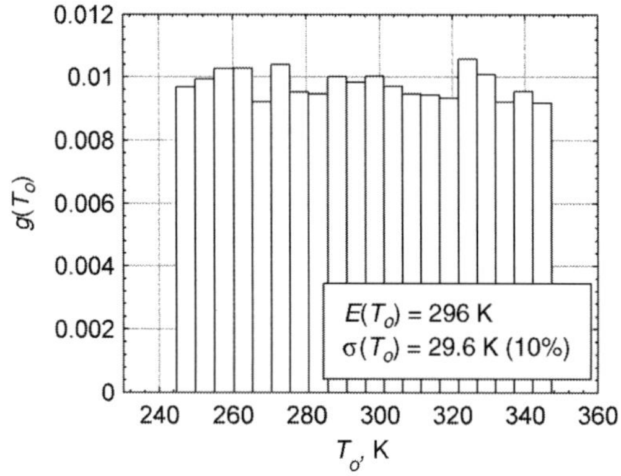

Figure 5.2 Probability density function of variable representing ambient temperature T_o (uniform distribution)

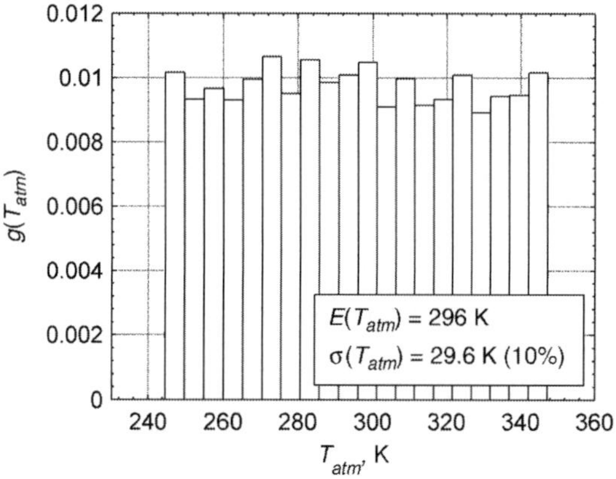

Figure 5.3 Probability density function of variable representing atmospheric temperature T_{atm} (uniform distribution)

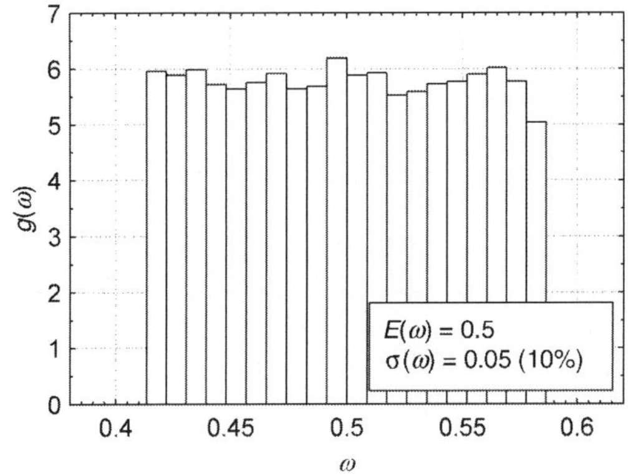

Figure 5.4 Probability density function of variable representing relative humidity ω (uniform distribution)

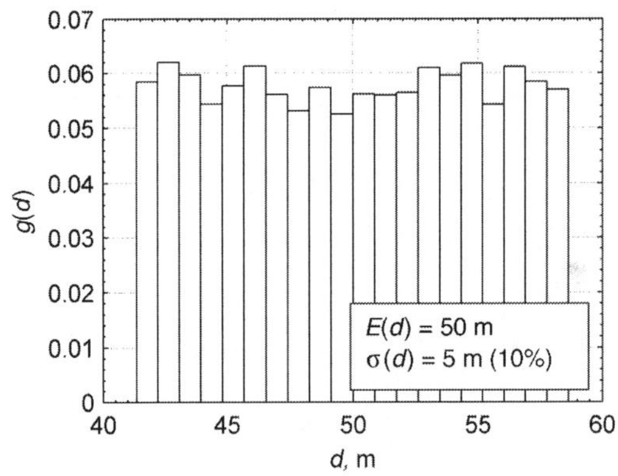

Figure 5.5 Probability density function of variable representing camera-to-object distance d (uniform distribution)

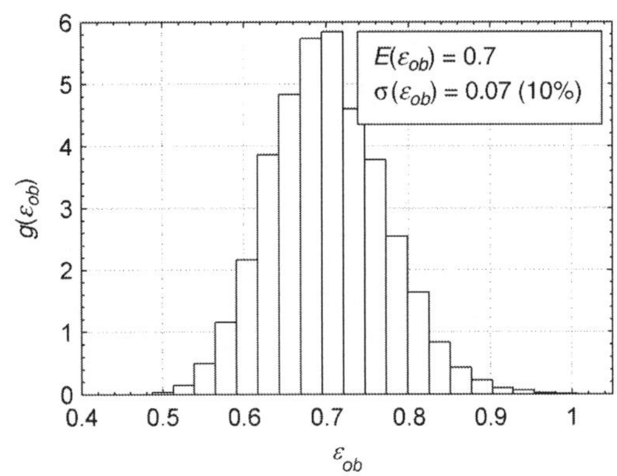

Figure 5.6 Probability density function of variable representing emissivity ε_{ob} (logarithmic Gaussian distribution)

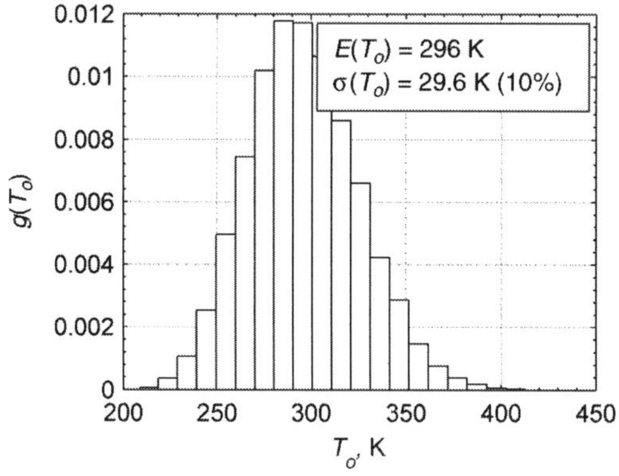

Figure 5.7 Probability density function of variable representing ambient temperature T_o (logarithmic Gaussian distribution)

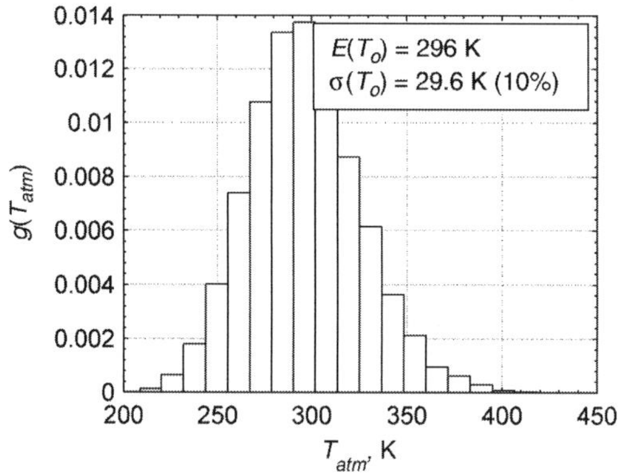

Figure 5.8 Probability density function of variable representing atmospheric temperature T_{atm} (logarithmic Gaussian distribution)

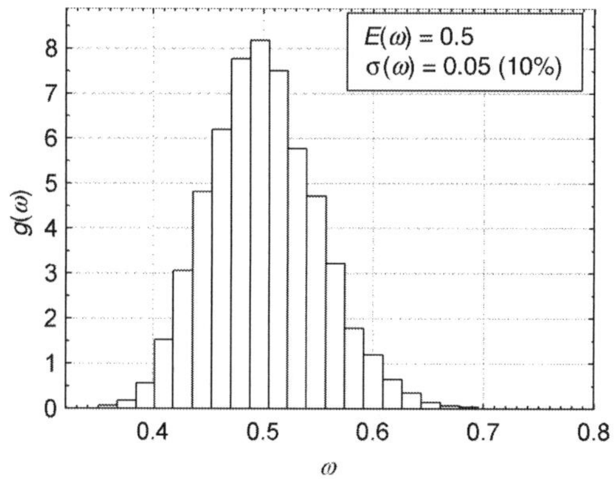

Figure 5.9 Probability density function of variable representing relative humidity ω (logarithmic Gaussian distribution)

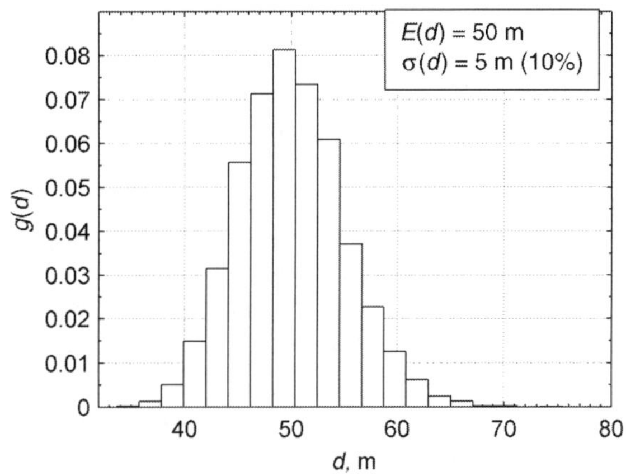

Figure 5.10 Probability density function of variable representing camera-to-object distance d (logarithmic Gaussian distribution)

5.2.4 Correlated Input Variables

Simulations of components of the combined standard uncertainty were conducted using MATLAB and its Statistic Toolbox (MATLAB 2005a). Consequently, the description of the generation of the input variables presented below refers to the MATLAB environment. Function `mvrnd()` allows for straightforward generation of multidimensional random variables of Gaussian marginal distributions. There is also the option of defining a covariance matrix for the generated variables. As a result, the variables are subject to the Gaussian distribution and are cross-correlated with defined correlation coefficients. Unfortunately, MATLAB's Statistic Toolbox does not include an equivalent of `mvrnd()` for the generation of multidimensional random variables of any marginal distribution shape. In MATLAB (2005b) there is a method for the generation of correlated variables of practically any (implemented in the Statistic Toolbox) marginal distribution. That method is used in this monograph to generate cross-correlated input variables of measurement model (3.17) and the atmospheric transmission model (3.15). The algorithm for this method can be divided into the following steps (Dudzik 2005):

1. Generation of the required number of pairs of two Gaussian random variables. These variables are correlated at levels specified by the corresponding entries of the covariance matrix.
2. Application of the Gaussian cumulative distribution function (CDF), denoted here by ϕ, to normalized Gaussian random variable Z. As a result, we obtain random variable U subject to the normalized uniform distribution on interval [0, 1]. The CDF of variable $U = \phi(Z)$ is expressed as (MATLAB 2005b):

$$Pr\{U \leq u_0\} = Pr\{\phi(Z) \leq u_0\} = Pr\{Z \leq \phi^{-1}(u_0)\} = u_0. \tag{5.11}$$

It is the CDF of uniform random variable U on interval [0, 1].

3. According to the theory of pseudorandom generators of one-dimensional random variables, application of the inverse CDF of any probability distribution F to (normalized uniform) random variable U yields a random variable subject to a distribution identical to F. The proof of this statement is the inverse of (5.11).

Hence, the correlating algorithm can generate realizations of a random variable of given distribution F. If the operation of generation is repeated for both original random variables, the resulting output variables, which inherit the interdependencies of the original variables, will have defined probability distributions. Unfortunately, application of nonlinear inverse CDFs changes the original cross-correlations of the variables: the coefficients of linear correlation of the resulting variables differ from the correlation coefficients of the original variables. For variables correlated in this way, it is advisable to use rank correlations (i.e. coefficients that define levels of nonlinear relationships between variables). In MATLAB (2005b), Spearman's rank correlation ρ and Kendall's rank correlation τ are used to assess nonlinear correlations between random variables. In this monograph we use the uniform distributions as the resulting distributions of correlated variables. In such a case, the linear correlation coefficient is retained, so it is further used as a numerical index of relationships between the variables.

The algorithm for the generation of the correlated input variables of model (3.17), with the atmospheric transmission model (3.15), was implemented in the MATLAB environment. The main window of the program is shown in Figure 5.11.

The program has the following functions:

- reading of measurement conditions from a thermogram file;
- generation of series of input variable realizations with a specified correlation coefficient;

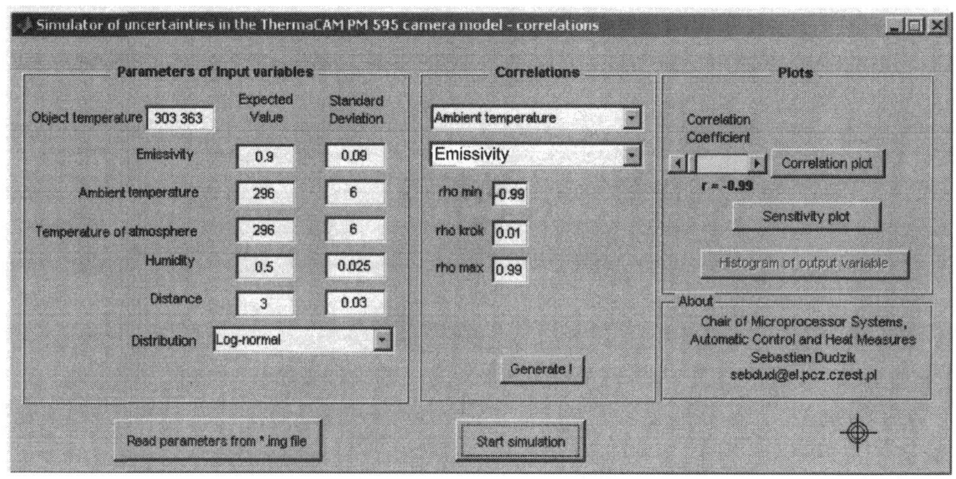

Figure 5.11 Main window of the program for investigating the influence of cross-correlations between the input quantities of the ThermaCAM PM 595 camera measurement model on the combined standard uncertainty (Dudzik 2007). See Color Plate 10 for the color version

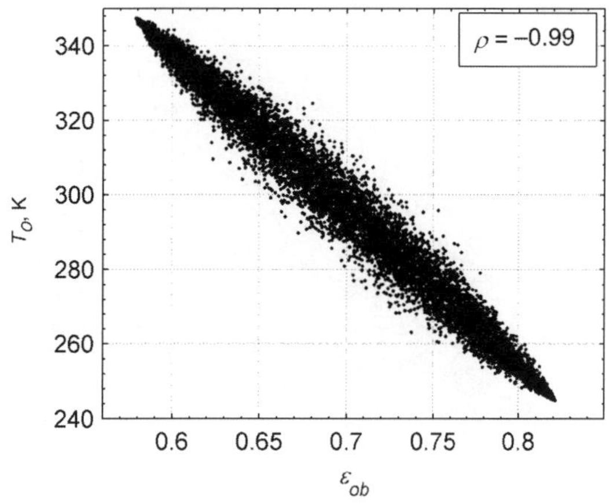

Figure 5.12 Simulation of the correlating algorithm for variables ε_{ob} and T_o, correlation $\rho = -0.99$ (uniform distributions)

- graphical presentation of the results, including:
  - cross-correlations,
  - sensitivity of the combined standard uncertainty of the model to changes in the input variable correlation coefficients, and
  - histograms of the input variables.

Selected results obtained from the program are presented below for illustrative purposes. Figures 5.12–5.16 show cross-correlations of two input variables: emissivity ε_{ob} and ambient temperature T_o, provided that they are subject to uniform probability distributions. Expected value $E(\varepsilon_{ob}) = 0.7$ and standard deviation $\sigma(\varepsilon_{ob}) = 0.07$ (10%) were assigned to the random

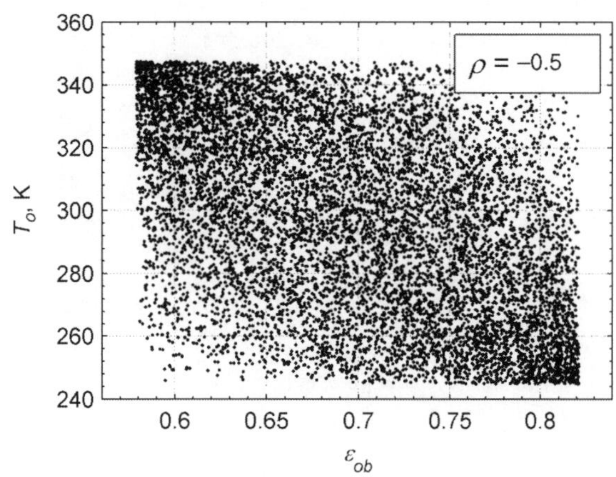

Figure 5.13 Simulation of the correlating algorithm for variables ε_{ob} and T_o, correlation $\rho = -0.5$ (uniform distributions)

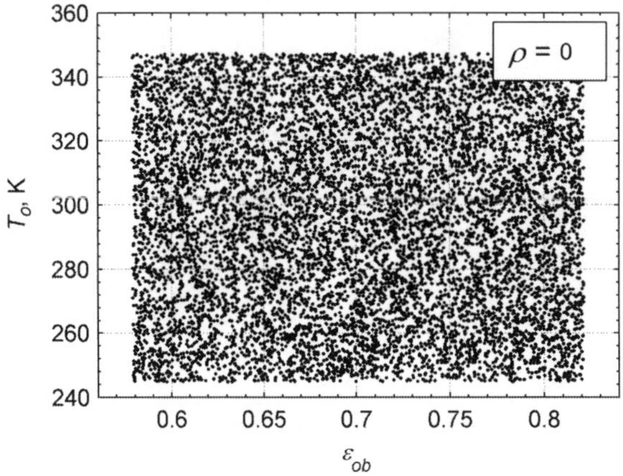

Figure 5.14 Simulation of the correlating algorithm for variables ε_{ob} and T_o, correlation $\rho = 0.0$ (uncorrelated variables, uniform distributions)

variable representing emissivity ε_{ob}, expected value $E(T_o) = 296\,\text{K}$ and standard deviation $\sigma(T_o) = 29.6\,\text{K}$ (10%) were assigned to the random variable representing ambient temperature T_o. Simulations were performed for five values of correlation coefficient ρ: –0.99, –0.50, 0, 0.50 and 0.99.

As mentioned earlier, the input variables of the considered measurement model were also generated using the logarithmic Gaussian distribution, and the correlating algorithm was tested for this distribution as well. Simulation results presented in Figures 5.17–5.21 were obtained for the same values of expected value E, standard deviation σ and correlation ρ as those for the uniform distribution (shown in Figures 5.12–5.16).

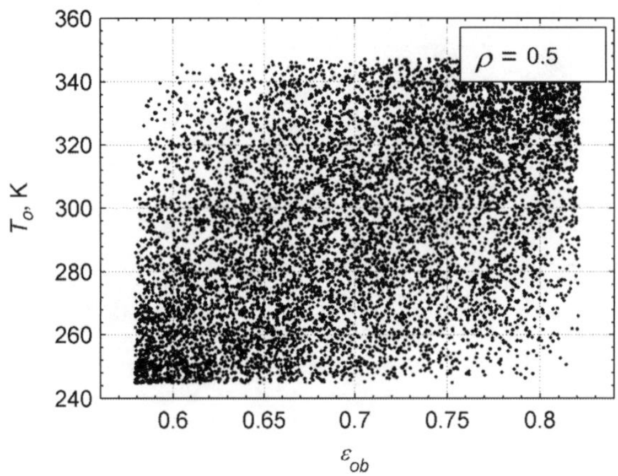

Figure 5.15 Simulation of the correlating algorithm for variables ε_{ob} and T_o, correlation $\rho = 0.5$ (uniform distributions)

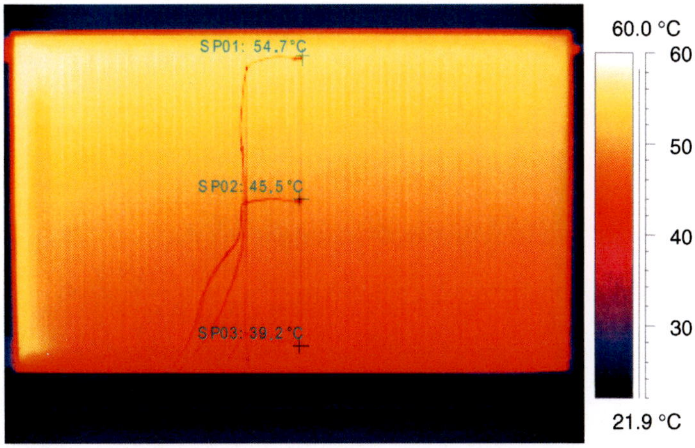

Plate 1 Thermogram of the heater with marked temperature measurement points. (See page 26)

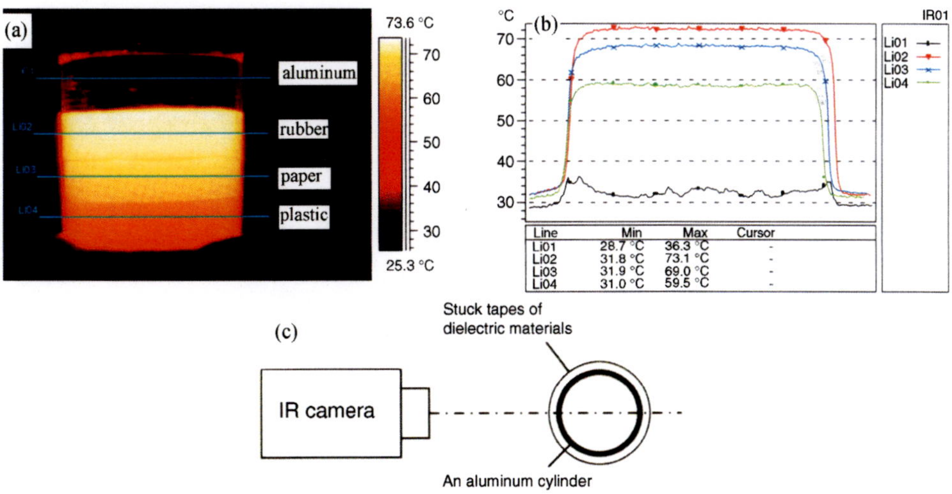

Plate 2 Infrared thermography temperature measurement of an aluminum cylinder (cross-section Li01) with stuck tapes of dielectric materials: rubber (cross-section Li02), paper (cross-section Li03) and plastic (cross-section Li04) under stationary conditions. Infrared camera shows an apparently different temperature for each material: (a) thermogram; (b) temperature profiles; and (c) top view of the experimental setup (Minkina 2004). Reproduced by permission of Częstochowa University of Technology. (See page 29)

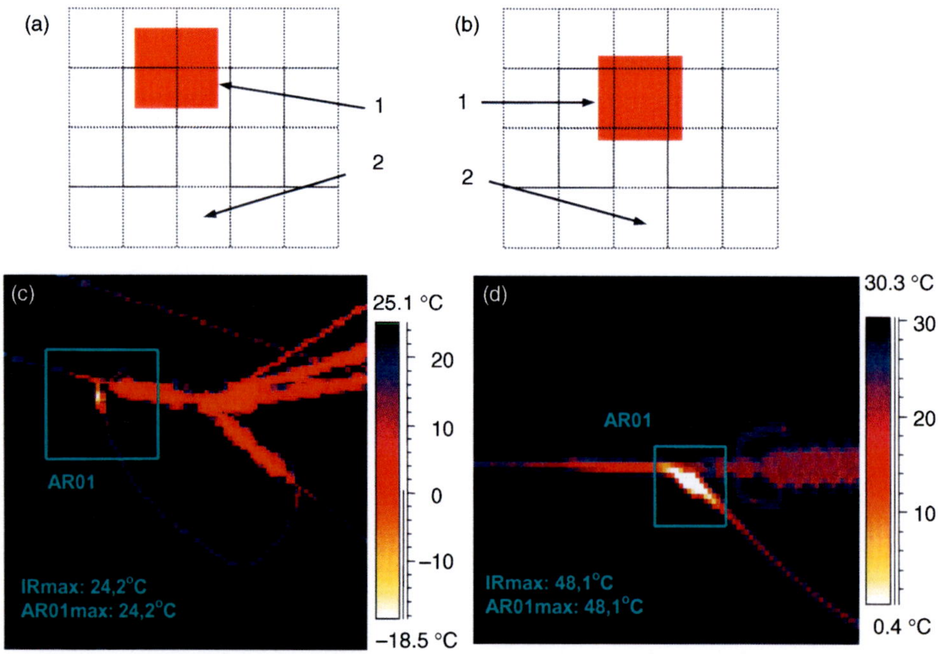

Plate 3 Image of a small object (anchor clamp for bridge connection of high-voltage line anchor support) on an array detector allowing correct temperature measurement: (a) no single detector is fully irradiated; (b) at least one detector is fully irradiated (1, object image; 2, array detectors); (c) thermogram of a clamp recorded at long distance – about 40 meters (optical and digital zoom); (d) thermogram of a clamp recorded at short distance – about 7 meters (only digital zoom) (Minkina 2001). (See page 37)

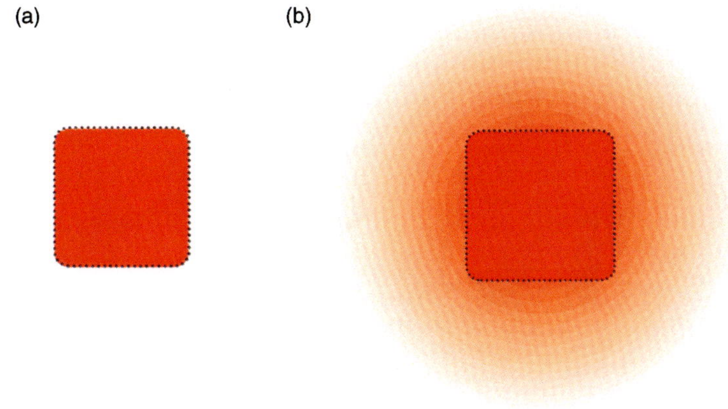

Plate 4 Determination of measurement area size: (a) ideal optics – irradiation of area of 2 × 2 detectors required for correct measurement; (b) real optics, image blurring – irradiation of area of 3 × 3 or 4 × 4 (sometimes 5 × 5) detectors required for correct measurement (Danjoux 2001, Minkina 2004). (See page 38)

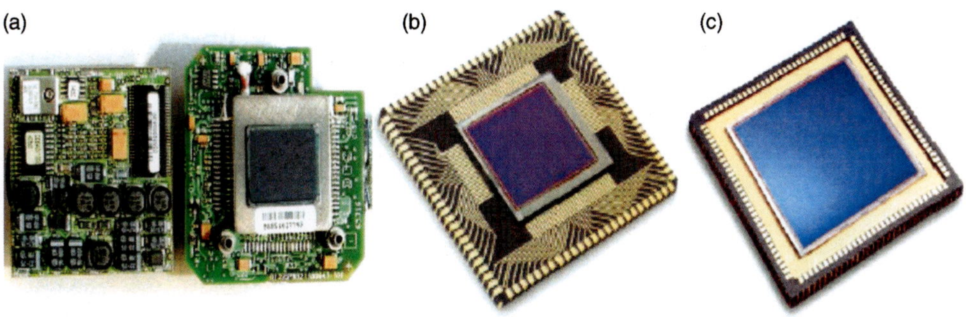

Plate 5 Array detectorsr (FPA) used in cameras from FLIR: (a) non-cooled microbolometer with thermoelectric stabilization (Peltier element), operating temperature about + 30 °C (www.flir.com, www.flir.com.pl, www.infraredtraining.com); and Raytheon (www.raytheon.com): (b) 512 × 512 ALADDIN III Quadrant; (c) 1024 × 1024 ALADDIN III. (See page 47)

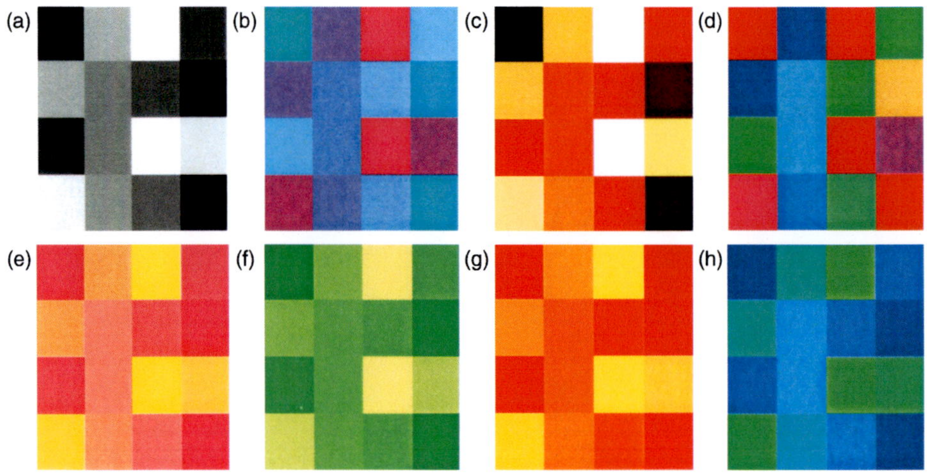

Plate 6 Enlarged digital image for different color maps: (a) grayscale; (b) cool; (c) hot; (d) HSI (Hue Saturation Intensity); (e) spring; (f) summer; (g) autumn; (h) winter. (See page 50)

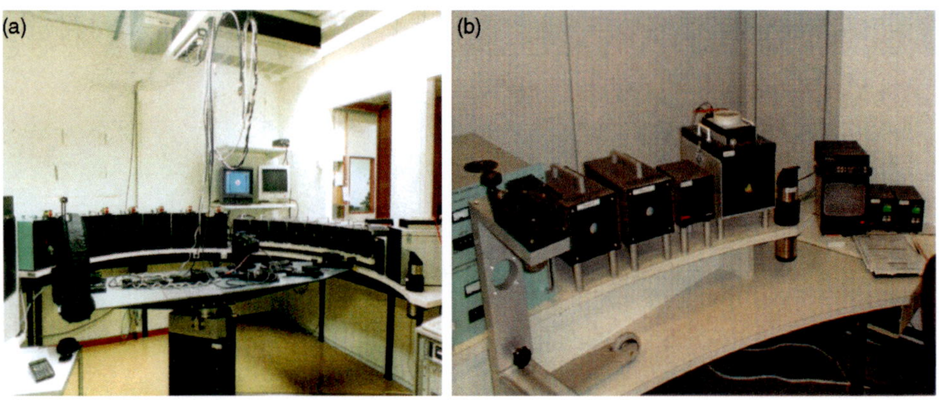

Plate 7 (a) Laboratory room for calibration of infrared cameras; (b) set of technical black bodies (IR-Book2 2000). Reproduced by permission of ITC Flir Systems. (See page 66)

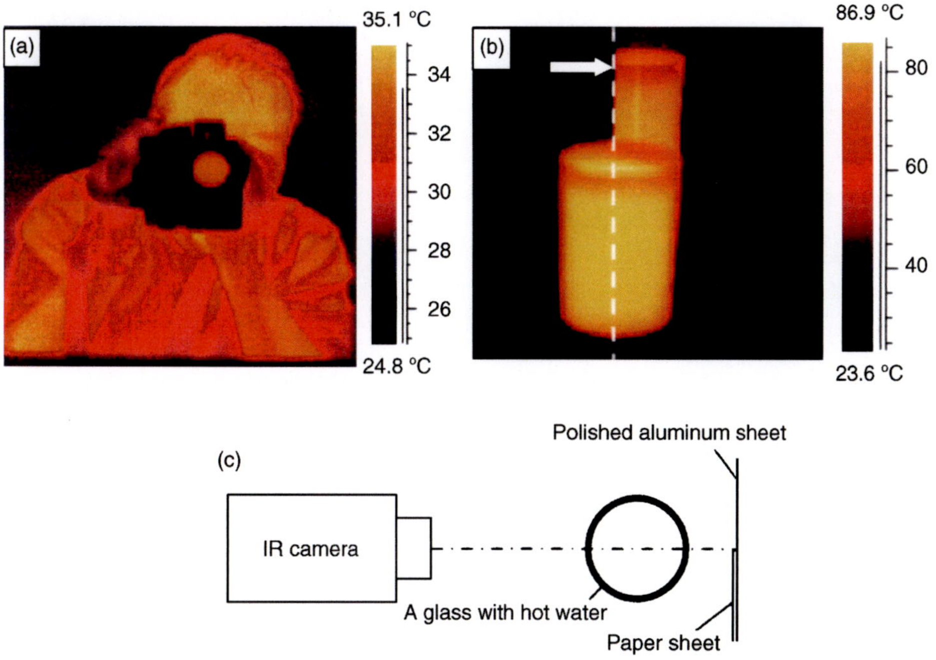

Plate 8 Thermograms of a polished aluminum sheet, whose temperature is close to the ambient temperature: (a) mirror reflection of a person measuring the sheet temperature; (b) image of a glass of hot water located against the sheet and its background reflection (right-hand edge of paper sheet stuck on the aluminum is marked with dashed line), (c) view (Minkina 2004). Reproduced by permission of Częstochowa University of Technology. (See page 71)

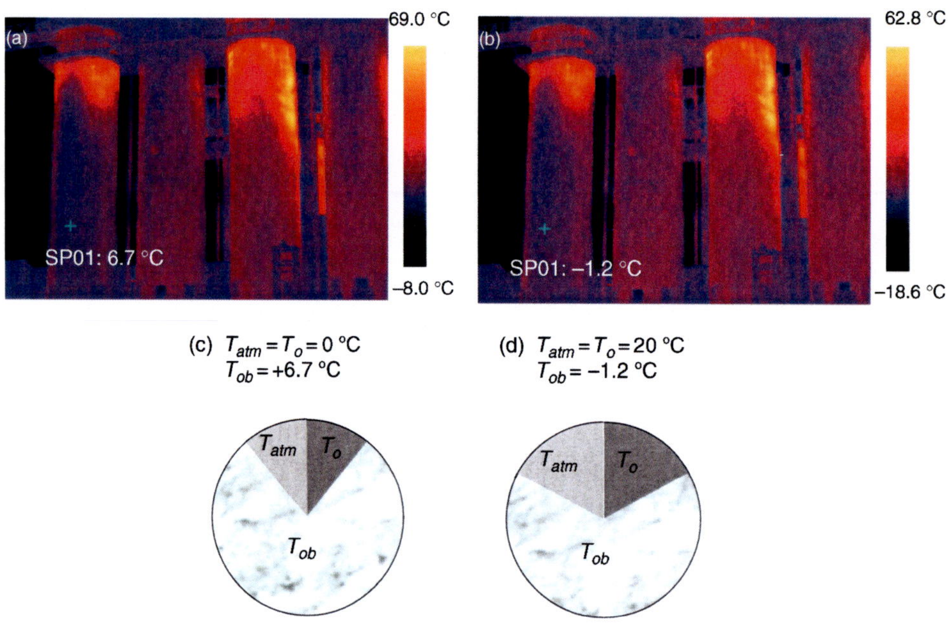

Plate 9 Influence of incorrect settings of T_{atm} and T_o on temperature measurement of an industrial installation (Minkina 2003). Reproduced by permission of Częstochowa University of Technology. (See page 76)

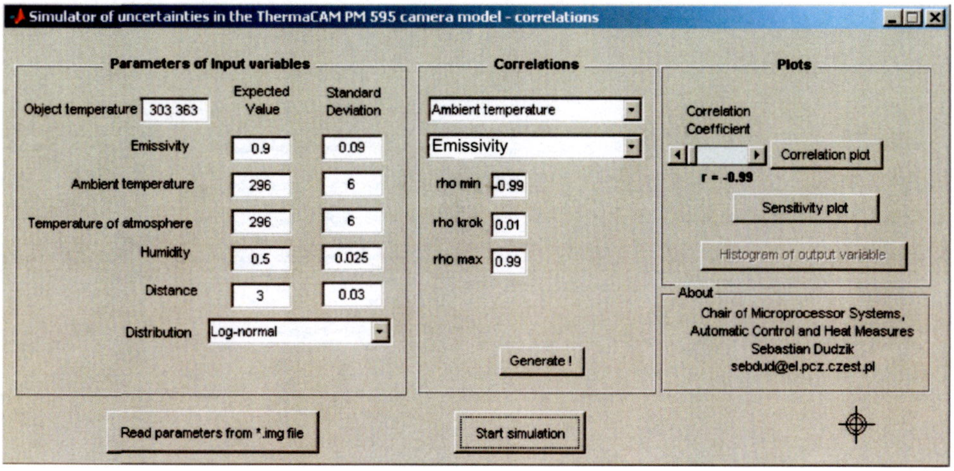

Plate 10 Main window of the program for investigating the influence of cross-correlations between the input quantities of the ThermaCAM PM 595 camera measurement model on the combined standard uncertainty (Dudzik 2007). (See page 90)

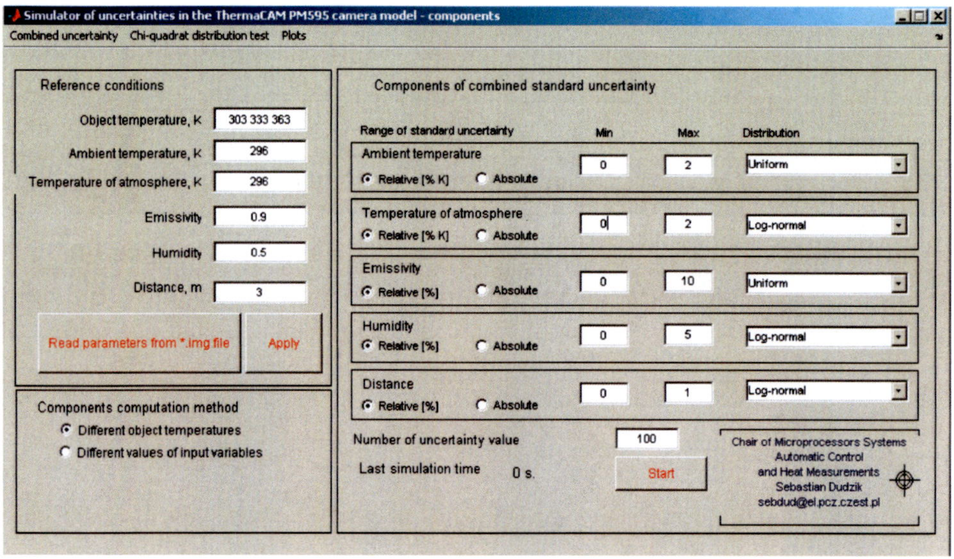

Plate 11 Main window of the program for simulation of sensitivity of the ThermaCAM PM 595 camera measurement model. (See page 96)

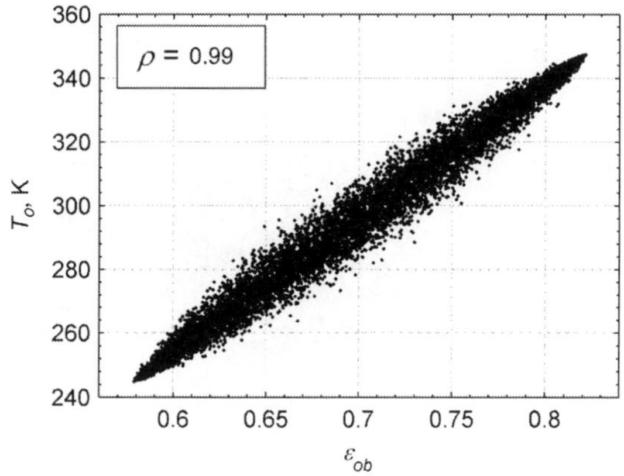

Figure 5.16 Simulation of the correlating algorithm for variables ε_{ob} and T_o, correlation $\rho = 0.99$ (uniform distributions)

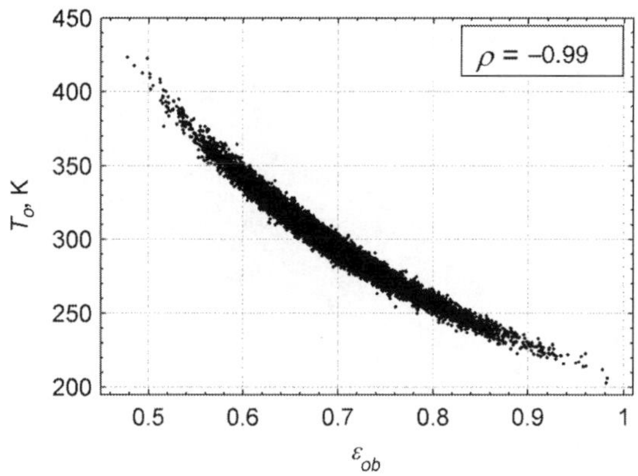

Figure 5.17 Simulation of the correlating algorithm for variables ε_{ob} and T_o, correlation $\rho = -0.99$ (logarithmic Gaussian distributions)

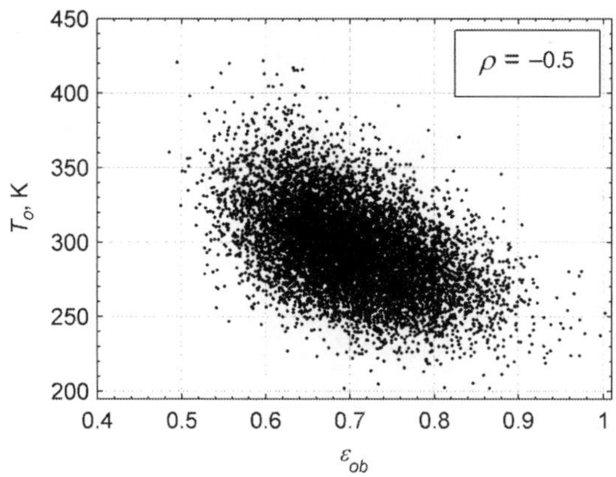

Figure 5.18 Simulation of the correlating algorithm for variables ε_{ob} and T_o, correlation $\rho = -0.5$ (logarithmic Gaussian distributions)

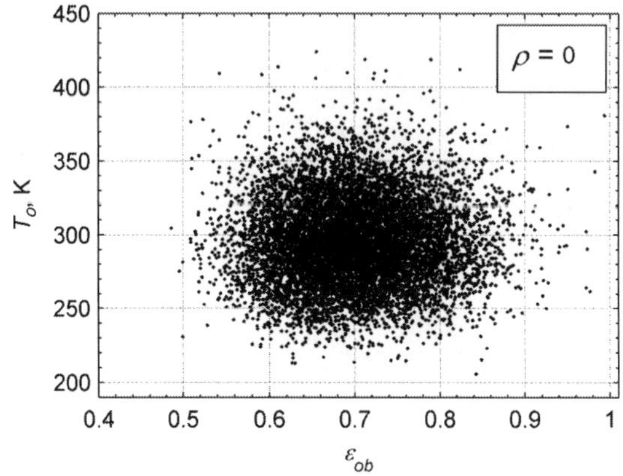

Figure 5.19 Simulation of the correlating algorithm for variables ε_{ob} and T_o, correlation $\rho = 0.0$ (uncorrelated variables, logarithmic Gaussian distributions)

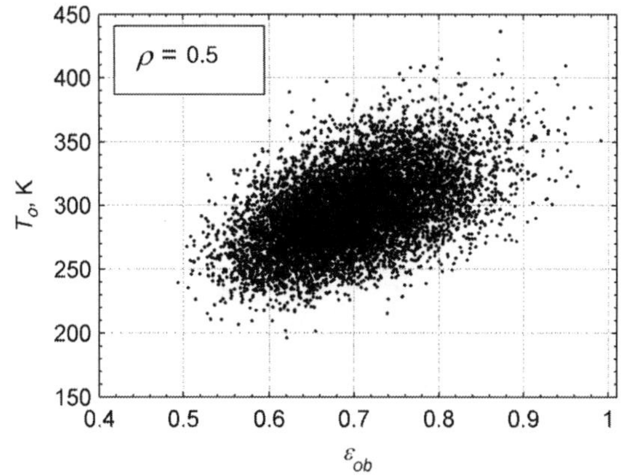

Figure 5.20 Simulation of the correlating algorithm for variables ε_{ob} and T_o, correlation $\rho = 0.5$ (logarithmic Gaussian distributions)

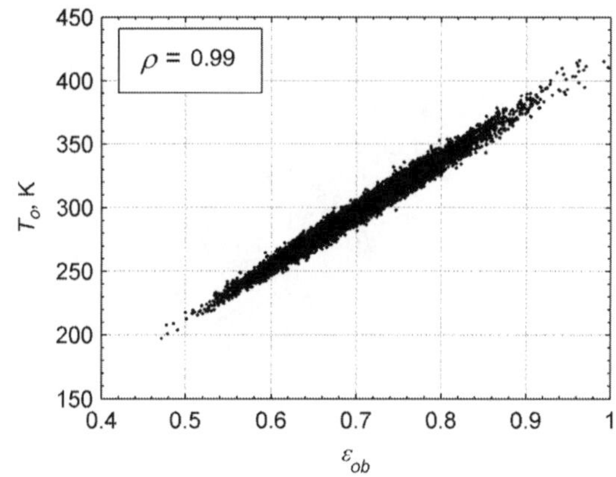

Figure 5.21 Simulation of the correlating algorithm for variables ε_{ob} and T_o, correlation $\rho = 0.99$ (logarithmic Gaussian distributions)

The influence of correlations between pairs of the model input variables on the combined standard uncertainty of temperature evaluation by means of infrared thermography, investigated using the algorithm described above, is discussed in Section 5.4. In Section 5.3 we evaluate and discuss components of the combined standard uncertainty assuming that the model input variables are uncorrelated.

5.3 Components of the Combined Standard Uncertainty for Uncorrelated Input Variables

Simulations of components of the combined uncertainty were conducted using the authors' software developed at the Electrical Faculty of the Częstochowa University of Technology. The software was written using MATLAB 7.1 (R13 SP1). MATLAB's built-in functions allow the generation of random variables representing the model input variables. The simulations presented in this section deal with the case of uncorrelated input variables.

The most important functions of the computer program are as follows:

- Generation of series of realizations with user-defined frequency distributions and parameters–Section 5.2.
- Reading reference values and calibration parameters from a thermogram AFF (AGEMA File Format) file (TOOLKIT IC2), and the atmospheric transmission model (3.15).
- Simulation of processing algorithm based on model (3.17).
- Graphical presentation of simulation results, including:
  - o frequency histograms of the input quantities;
  - o frequency histograms of components of the combined standard uncertainty;
  - o linear graphs of components of the combined standard uncertainty.

The main window of the program for simulations of the infrared camera measurement model is shown in Figure 5.22.

The frequency distributions of the variables are determined by series of realizations generated for user-defined parameters (Gajda and Szyper 1998). For simulation of the uncertainty components the two probability distributions described in Section 5.2 were employed: the uniform and the logarithmic Gaussian distribution. In this monograph we investigate the influence of probability distributions of the five input variables of measurement model (3.17) and the atmospheric transmission model (3.15) on the probability distribution function of the model output variable. The aim of the simulations was to evaluate uncertainty components of the processing algorithm associated with the influence of a particular input on the joint uncertainty.

As for the case of simulation of errors in Chapter 4, the analysis of uncertainties was carried out for four values of the object temperature: 30 °C (303 K), 50 °C (323 K), 70 °C (343 K) and 90 °C (363 K). The settings of reference values are as in Table 5.2. The ranges of the input variable uncertainty assumed for simulations are given in Table 5.3. To investigate the dependence of particular components of the combined standard uncertainty on emissivity ε_{ob} and camera-to-object distance d, the processing algorithm was simulated for four different values of ε_{ob} and two values of d.

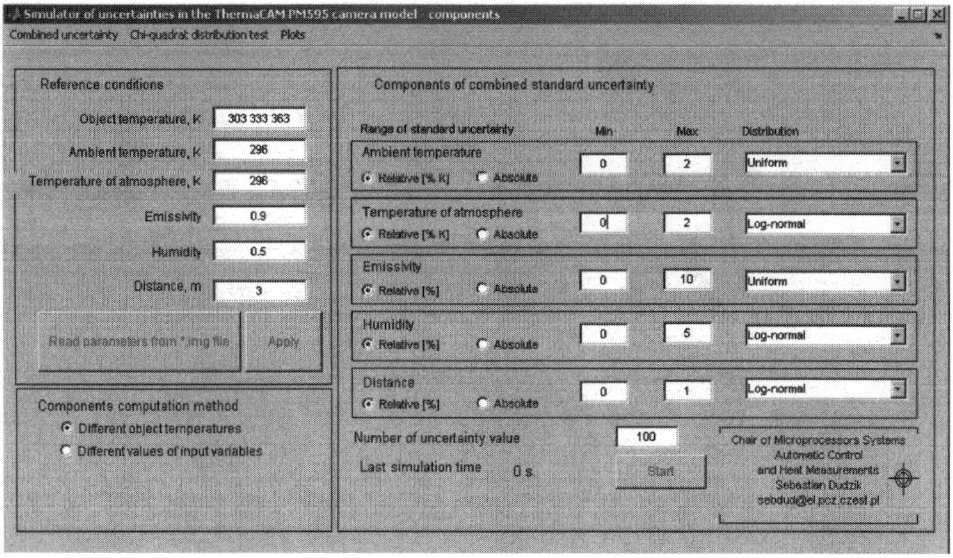

Figure 5.22 Main window of the program for simulation of sensitivity of the ThermaCAM PM 595 camera measurement model. See Color Plate 11 for the color version

5.3.1 Component of the Combined Standard Uncertainty Associated with the Object Emissivity

Figures 5.23 and 5.24 show the simulation results for the uncertainty component associated with uncertainty $u(\varepsilon_{ob})$ of the object emissivity, assuming a uniform probability distribution. The simulations were conducted for four values of the object emissivity ε_{ob} and two values of the camera-to-object-distance d as given in Table 5.2.

Simulation results for the uncertainty component associated with uncertainty $u(\varepsilon_{ob})$ of the object emissivity, assuming $d = 100$ m, are shown in Figure 5.24.

Analysis of the graphs in Figures 5.23 and 5.24 allows the following conclusions to be drawn:

- The component $u(T_{ob})$ of the combined standard uncertainty associated with emissivity ε_{ob} depends strongly on object temperature T_{ob}. For example, from Figure 5.23a we can see that an increase in the object temperature by 60 K (from 303 to 363 K) results in a five-times increase in the relative combined standard uncertainty (from about 1% to over 5%) for the maximum considered standard uncertainty of the object emissivity $u(\varepsilon_{ob}) = 30\%$.
- An assumed value of the object emissivity does not affect the considered component of the combined standard uncertainty. For example, Figures 5.232a–d show that for object temperature $T_{ob} = 323$ K, the value of the component is the same, equal to $u(T_{ob}) = 3\%$ for standard uncertainty $u(\varepsilon_{ob}) = 30\%$.
- The component $u(T_{ob})$ of the combined standard uncertainty associated with emissivity ε_{ob} does not depend on an assumed camera-to-object distance d. In fact, comparing Figures 5.23 and 5.24, we can see that the graphs of the object temperature T_{ob} relative to the uncertainty associated with emissivity ε_{ob} are identical in both figures, regardless of object temperature T_{ob}.

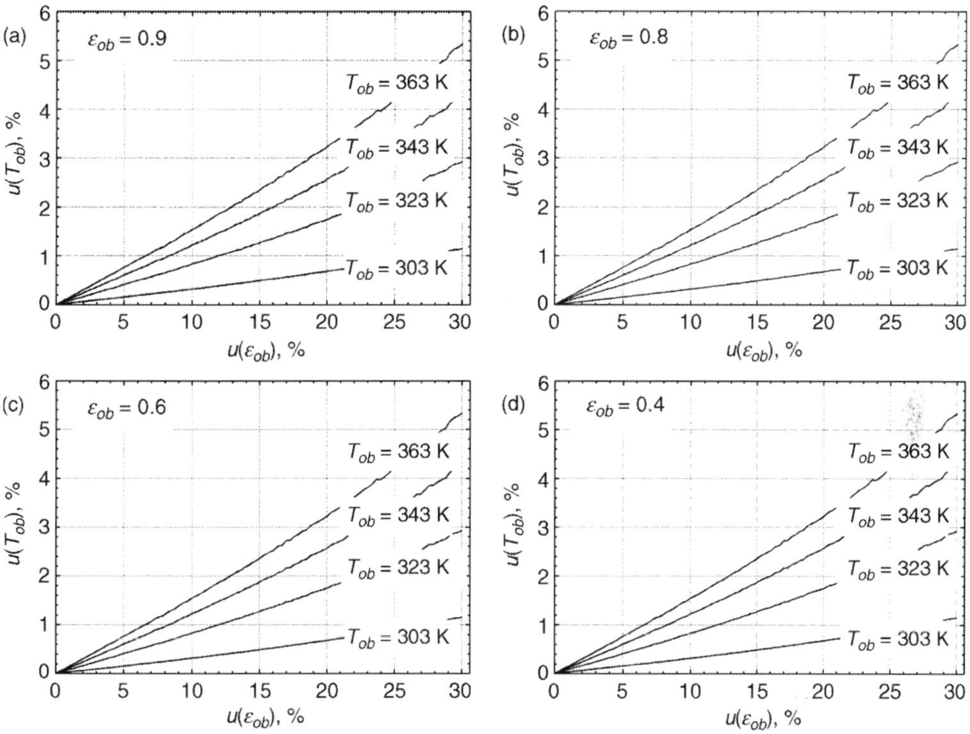

Figure 5.23 Component of the relative standard uncertainty associated with emissivity ε_{ob}, assuming a uniform distribution. Results for camera-to-object distance $d = 1$ m

- Comparing the simulation results for the component $u(T_{ob})$ of the object temperature, relative to the uncertainty associated with the object emissivity, to results obtained for the other components (presented further in this section), we can state that the standard uncertainty associated with the object emissivity has the strongest impact on the combined uncertainty of the temperature measurement for the model.

5.3.2 Component of the Combined Standard Uncertainty Associated with the Ambient Temperature

Simulations of the uncertainty component associated with uncertainty $u(T_o)$ of the ambient temperature were performed, as in the previous section, for different values of object emissivity and camera-to-object distance. The results for $d = 1$ m are shown in Figure 5.25 and the results for $d = 100$ m in Figure 5.26.

On the basis of the computation results presented in Figures 5.25 and 5.26, we can state that:

- The component $u(T_{ob})$ of the object temperature combined standard uncertainty associated with ambient temperature T_o depends strongly on the emissivity. Comparing

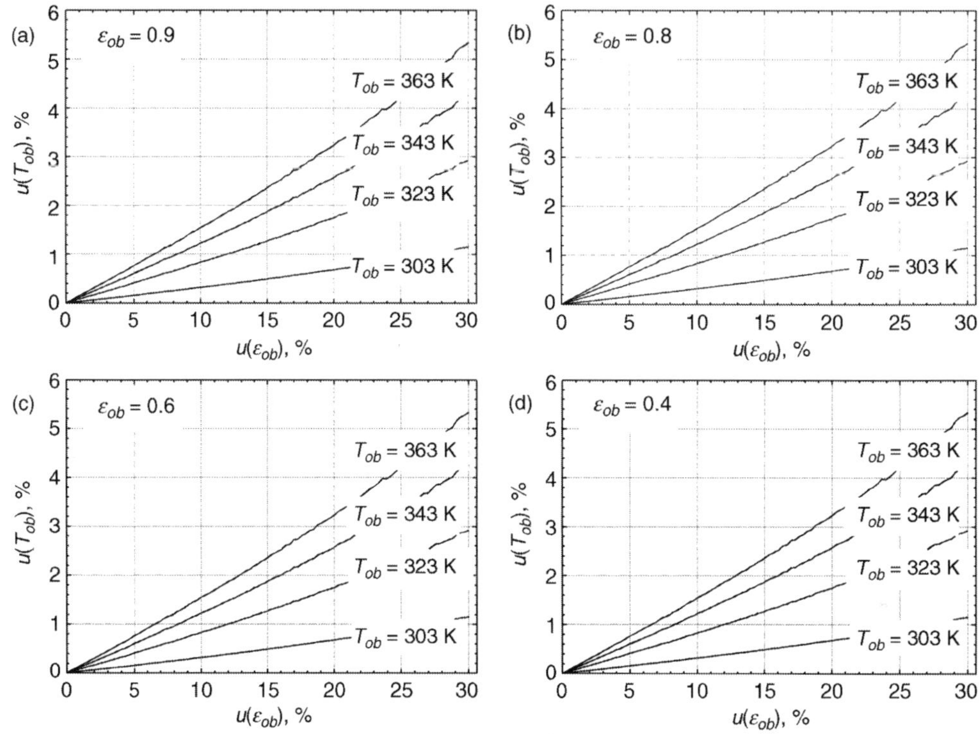

Figure 5.24 Component of the relative standard uncertainty associated with emissivity ε_{ob}, assuming a uniform distribution. Results for camera-to-object distance $d = 100$ m

graphs corresponding to object temperature $T_{ob} = 343$ K in Figures 5.25a–d, we can see that for $u(T_o) = 3\%$, the considered component changes from about 0.2% for $\varepsilon_{ob} = 0.9$ to about 2.6% for $\varepsilon_{ob} = 0.4$. The simulations show that this component decreases with the object emissivity.

- The component also depends on assumed object temperature T_{ob}. For example, looking at Figure 5.25d, we can see that $u(T_{ob})$ associated with T_o is equal to about 2% for uncertainty $u(T_o) = 3\%$ and temperature $T_{ob} = 363$ K and about 4% for $T_{ob} = 303$ K. In addition, all graphs in Figures 5.25 and 5.26 show that the higher the object temperature T_{ob}, the weaker the influence of the ambient temperature uncertainty $u(T_o)$ on the combined standard uncertainty.
- From the above observations we can conclude that the influence of the uncertainty of the ambient temperature on temperature measurement accuracy can be neglected for very high measured temperatures, especially if the object emissivity is high. This conclusion is *very important* for the practice of measurements in infrared thermography.
- Comparing Figures 5.25 and 5.26, we can state that the considered component of the relative combined standard uncertainty depends very little on the camera-to-object distance d.

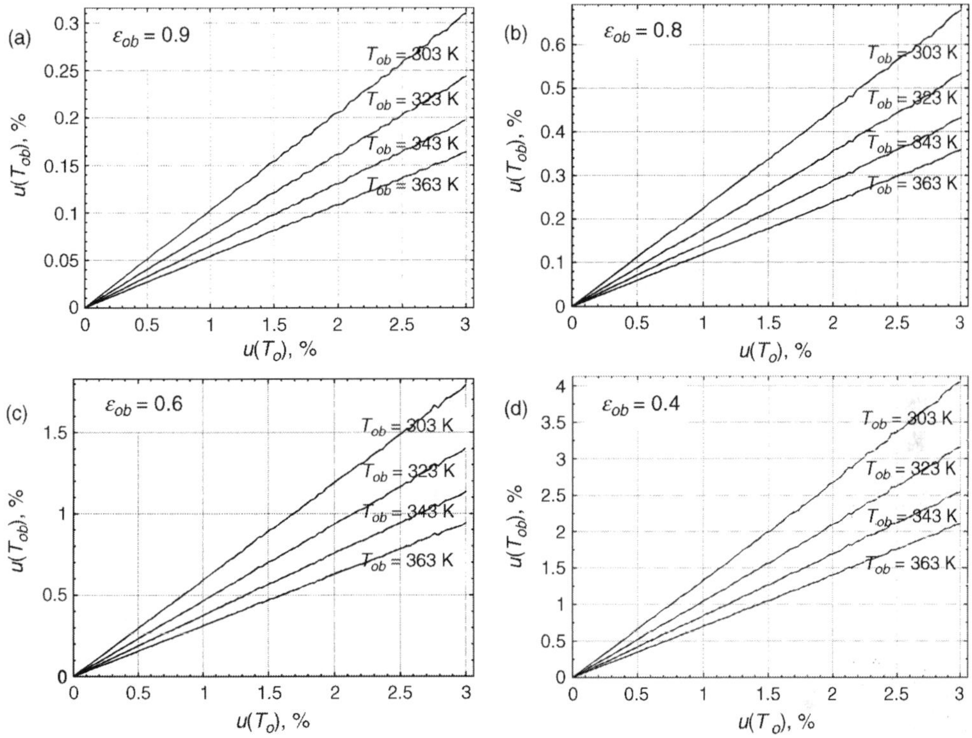

Figure 5.25 Component of the relative standard uncertainty associated with ambient temperature T_o, assuming a uniform distribution. Results for camera-to-object distance $d = 1$ m

5.3.3 *Component of the Combined Standard Uncertainty Associated with the Atmospheric Temperature*

Figures 5.27 and 5.28 show the simulation results for the uncertainty component associated with uncertainty $u(T_{atm})$ of the atmospheric temperature. The simulations were conducted under the same conditions as those defined in the previous sections.

Analysis of the graphs shown in Figures 5.27 and 5.28 allows the following conclusions to be made:

- Object emissivity ε_{ob} does not affect the component of the relative standard uncertainty associated with atmospheric temperature T_{atm} in Figures 5.27 and 5.28.
- Component $u(T_{ob})$ associated with the atmospheric temperature depends on object temperature T_{ob}. From Figure 5.28c we can see, for example, that for $u(T_{atm}) = 3\%$, it is somewhat above 0.05% when $T_{ob} = 303$ K and about 0.3% when $T_{ob} = 363$ K, that is nearly six times more.
- Comparison of results presented in Figures 5.27 and 5.28 indicates that the considered component of uncertainty depends on the camera-to-object distance d. For example, the value of $u(T_{ob})$ read from Figure 5.27b, for $u(T_{atm}) = 3\%$ and $T_{ob} = 323$ K, is somewhat

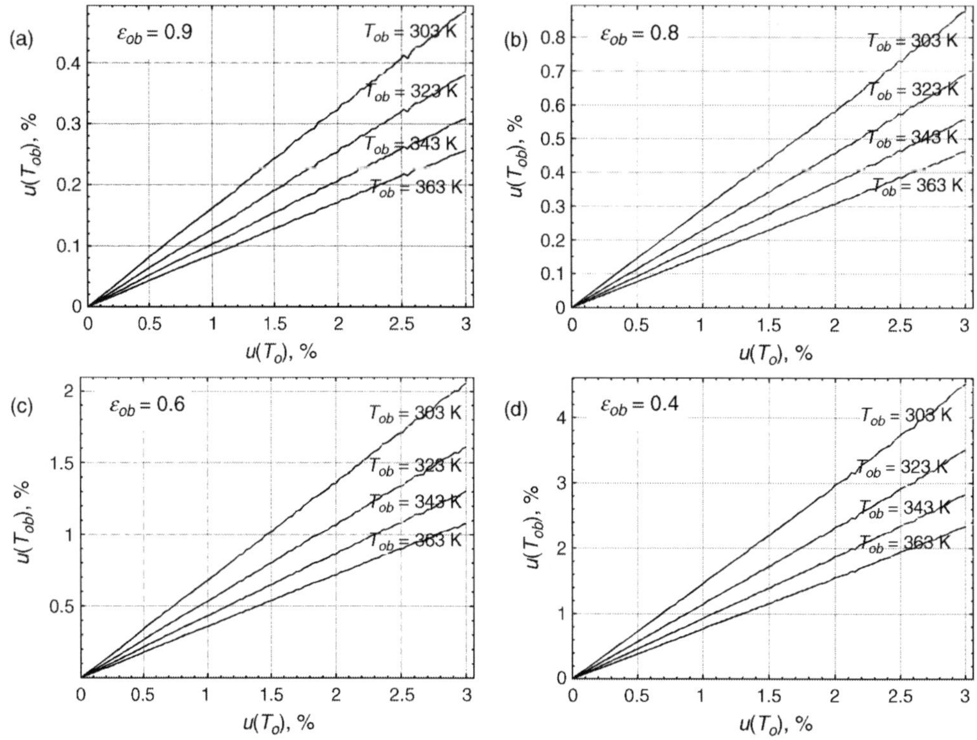

Figure 5.26 Component of the relative standard uncertainty associated with ambient temperature T_o, assuming a uniform distribution. Results for camera-to-object distance $d = 100$ m

above 0.01 %, whereas its value read from Figure 5.28b, for the same values of $u(T_{atm})$ and T_{ob}, is about 0.15%, that is 15 times more.

- Analyzing the simulations of uncertainty component $u(T_{ob})$ associated with T_{atm}, we can conclude that its contribution to the combined standard uncertainty of temperature measurement is practically negligible. Standard uncertainty $u(T_{atm})$ affects the combined uncertainty more visibly only for very long camera-to-object distance d.

5.3.4 Component of the Combined Standard Uncertainty Associated with the Atmospheric Relative Humidity

Simulation results for the combined uncertainty component associated with uncertainty $u(\omega)$ of the atmospheric relative humidity are shown in Figures 5.29 and 5.30. The simulations were performed, as before, for four different values of emissivity ε_{ob} and two values of camera-to-object distance d.

On the basis of the computation results presented in Figures 5.29 and 5.30, we can state that:

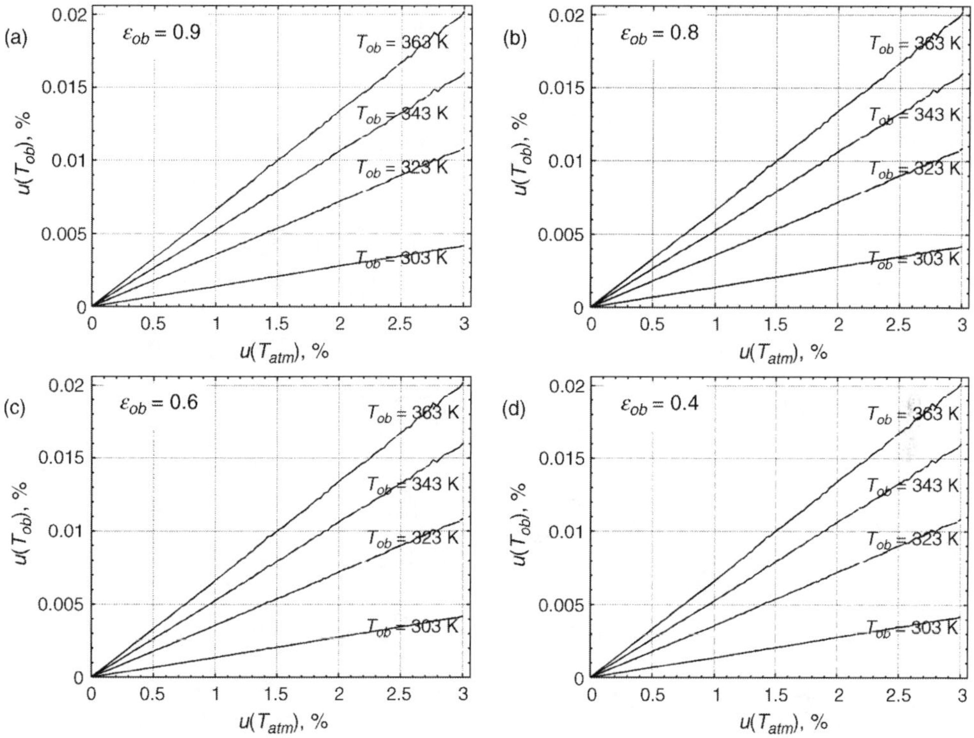

Figure 5.27 Component of the relative standard uncertainty associated with atmospheric temperature T_{atm}, assuming a uniform distribution. Results for camera-to-object distance $d = 1$ m

- Component $u(T_{ob})$ of the combined standard uncertainty associated with uncertainty $u(\omega)$ does not depend on object emissivity ε_{ob} – Figures 5.29 and 5.30. From Figure 5.30 we can see that the graphs of the considered component are identical in all four cases a–d. In fact, comparison of the results in Figure 5.30 for $u(\omega) = 30\%$ and $T_{ob} = 363$ K, for example, shows that in all four cases the value of the component is about 0.15%.
- The component of the combined uncertainty depends on object temperature T_{ob}, and its contribution to the total uncertainty of temperature measurement increases with T_{ob}. Both for $d = 1$ m (Figure 5.29) and for $d = 100$ m (Figure 5.30), the component increases with T_{ob} for any fixed uncertainty $u(\omega)$.
- Comparison of results presented in Figures 5.29 and 5.30 indicates that $u(T_{ob})$ associated with $u(\omega)$ depends on camera-to-object distance d. For example, the value of $u(T_{ob})$ read from Figure 5.29b for $u(\omega) = 30\%$ and $T_{ob} = 323$ K is about 0.006%, whereas the value of $u(T_{ob})$ read from Figure 5.30b, for the same values of $u(\omega)$ and T_{ob}, is about 0.08%.
- Analysis of the simulation of the relative standard uncertainty component associated with the relative humidity uncertainty $u(\omega)$ shows that it is negligible. The influence of this component on the combined temperature measurement uncertainty is even weaker than the influence associated with $u(T_{atm})$, discussed earlier.

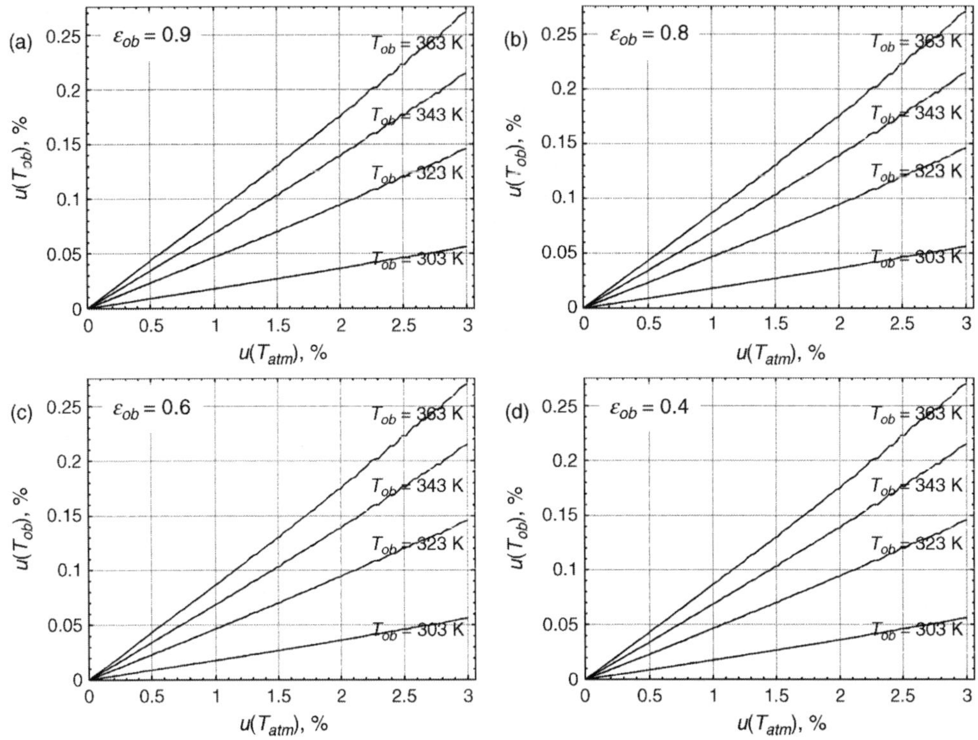

Figure 5.28 Component of the relative standard uncertainty associated with atmospheric temperature T_{atm}, assuming a uniform distribution. Results for camera-to-object distance $d = 100\,\text{m}$

5.3.5 Component of the Combined Standard Uncertainty Associated with the Camera-to-object Distance

Figures 5.31 and 5.32 show the simulation results for the uncertainty component associated with uncertainty $u(d)$ of the camera-to-object distance d. The simulations were carried out under the same conditions as those defined in the previous sections.

Analysis of the graphs shown in Figures 5.31 and 5.32 allows the following conclusions to be drawn:

- The value of the uncertainty component associated with the camera-to-object distance does not depend on the object emissivity assumed for simulations (as in the cases of components associated with ε_{ob}, T_{atm} and ω, discussed earlier). For example, looking at Figure 5.31 for $u(d) = 30\%$ and $T_{ob} = 363\,\text{K}$, we can see that $u(T_{ob}) \approx 0.014\%$ for all four cases a–d of ε_{ob}.
- The considered component depends on object temperature T_{ob}, and its contribution to the relative combined standard uncertainty increases with T_{ob}. In fact, taking into consideration

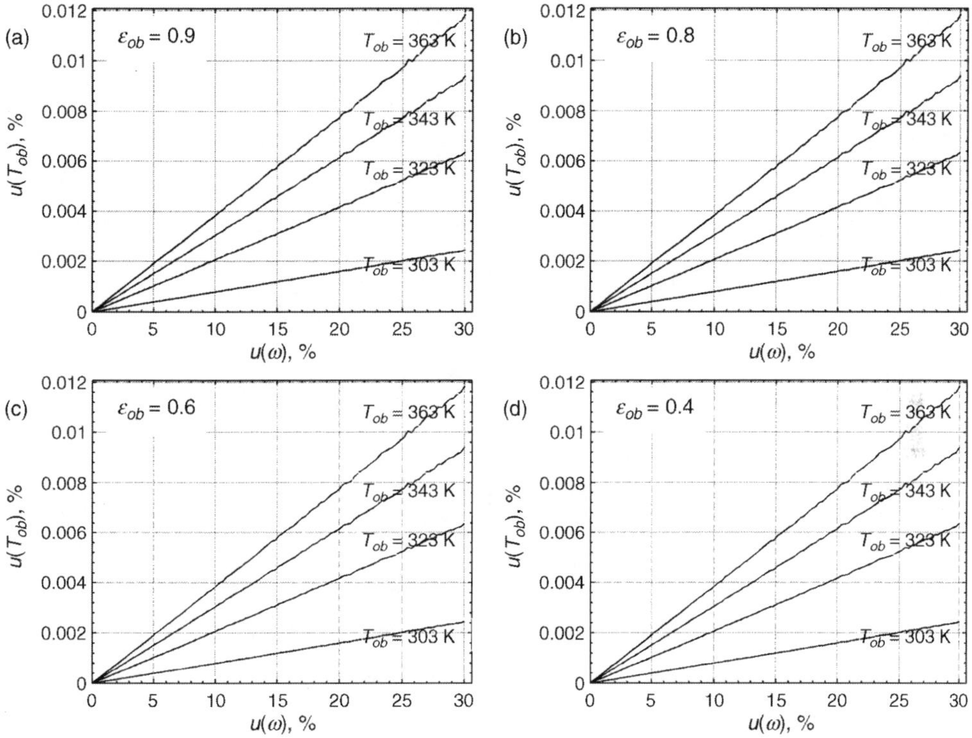

Figure 5.29 Component of the relative standard uncertainty associated with relative humidity ω, assuming a uniform distribution. Results for camera-to-object distance $d = 1$ m

the graphs from Figure 5.32a, it is easy to see that for $u(d) = 30\%$ and $T_{ob} = 303$ K, the uncertainty component associated with distance d is equal to about 0.03%, whereas for $T_{ob} = 363$ K, it exceeds 0.15%.

- Comparison of results shown in Figures 5.31 and 5.32 leads to the observation that the component depends on the camera-to-object distance d. For example, $u(T_{ob})$ associated with d read from Figure 5.31b for $u(T_{atm}) = 30\%$ and $T_{ob} = 323$ K is about 0.008%, whereas the corresponding $u(T_{ob})$ (for the same values of $u(T_{atm})$ and T_{ob}) read from Figure 5.32b is about 0.08%, that is 10 times more.
- Analysis of results presented in Figures 5.31 and 5.32 shows that the influence of the uncertainty component associated with distance d on the whole uncertainty budget can be neglected in practice, even for large uncertainty $u(d)$.

Comparing the analysis of errors, carried out in Chapter 4, to the analysis of uncertainties presented above allows us to conclude that in both cases components (of the total error or the combined uncertainty) associated with the same input quantity exhibit similar character.

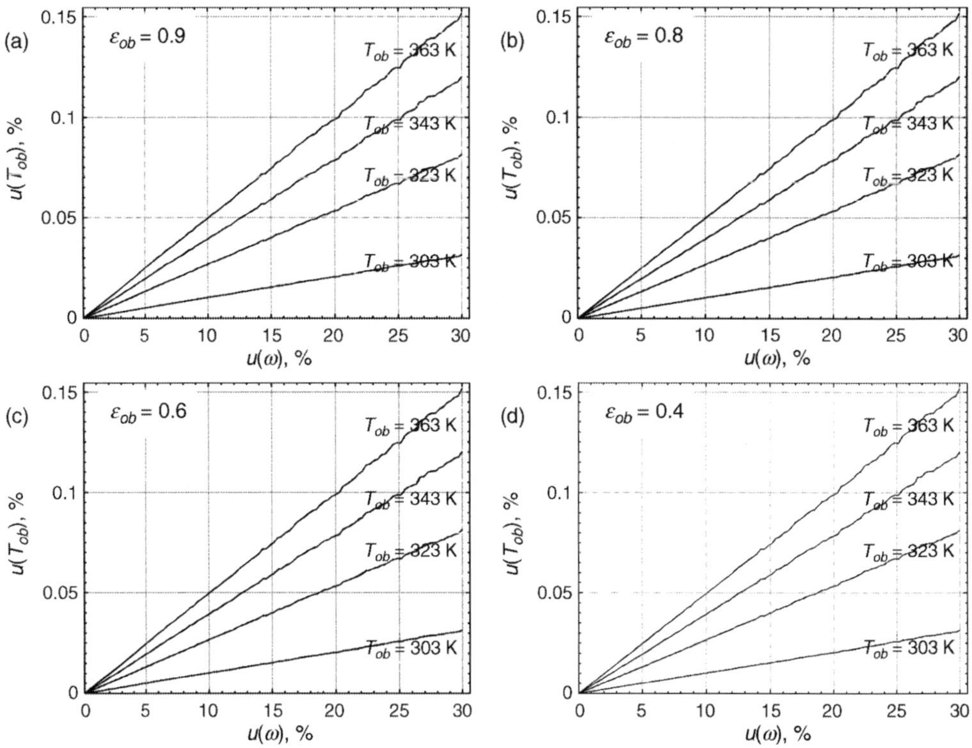

Figure 5.30 Component of the relative standard uncertainty associated with relative humidity ω, assuming a uniform distribution. Results for camera-to-object distance $d = 100$ m

5.4 Simulations of the Combined Standard Uncertainty for Correlated Input Variables

5.4.1 Introduction

In Section 5.3 we investigated the influence of particular input variables of models (3.17) and (3.15) (Figures 3.10b–d) on the relative combined standard uncertainty of infrared thermography measurement. Simulations conducted for the models allowed us to evaluate the components of the combined uncertainty, provided that the input variables are uncorrelated. In reality, measurements of two or more input quantities might be statistically correlated. The cross-correlation between two random variables is expressed qualitatively for $-1 \leq \rho \leq 1$, as defined by (Taylor 1997), as:

$$\rho = \frac{\sum (x_i - \bar{x})(y_i - \bar{y})}{\left[\sum (x_i - \bar{x})^2 \sum (y_i - \bar{y})^2\right]^{1/2}}. \tag{5.12}$$

In this section we are concerned with the influence of correlations between pairs of the input variables of models (3.17) and (3.15) (Figures 3.10b–d) on the relative combined standard uncertainty. Simulations of the models were based on the methodology described

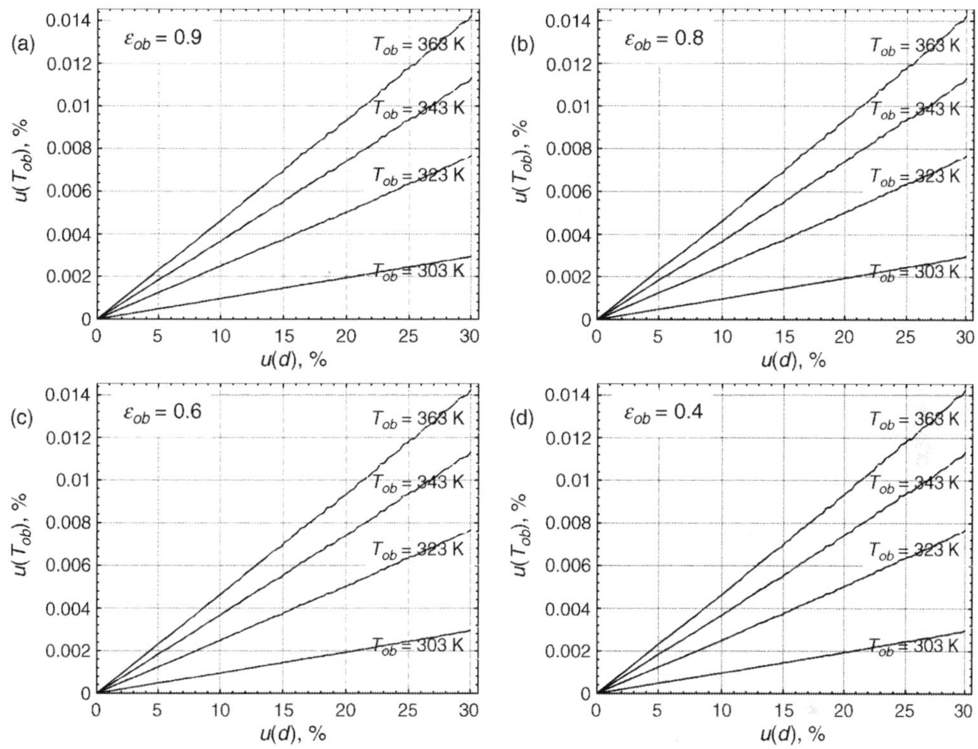

Figure 5.31 Component of the relative standard uncertainty associated with camera-to-object distance d, assuming a uniform distribution. Results for $d = 1$ m

in Section 5,1. Example distributions of random variables representing individual input quantities of the considered model, generated using the correlation algorithm, are shown in Figures 5.12–5.21. Presented below are the results of simulations carried out for each pair of input variables. The input data for simulations, namely assumed estimates and relative standard uncertainties of the input quantities, are given in Tables 5.4 and Table 5.5 respectively.

Analysis of the model in terms of dependence of the temperature measurement combined uncertainty on cross-correlations among the input variables was performed for three selected values of the object temperature: $T_{ob} = 323$ K (50 °C), $T_{ob} = 343$ K (70 °C) and $T_{ob} = 363$ K (90 °C). Since in reality accurate ambient temperature T_o is unknown, it is assumed to be equal to atmospheric temperature T_{atm} (as justified earlier). Therefore, as we can see from Table 5.5, the dependence of the combined uncertainty on cross-correlations among the input variables is investigated by taking into consideration the standard uncertainty contributed by distance d.

The situations considered in this monograph do not exhaust the issue of the dependence of the combined uncertainty on correlations among the model input variables. This is because this dependence is strongly affected by the measurement conditions (i.e. estimates of the influencing quantities, standard uncertainties of particular input variables, etc.). Other

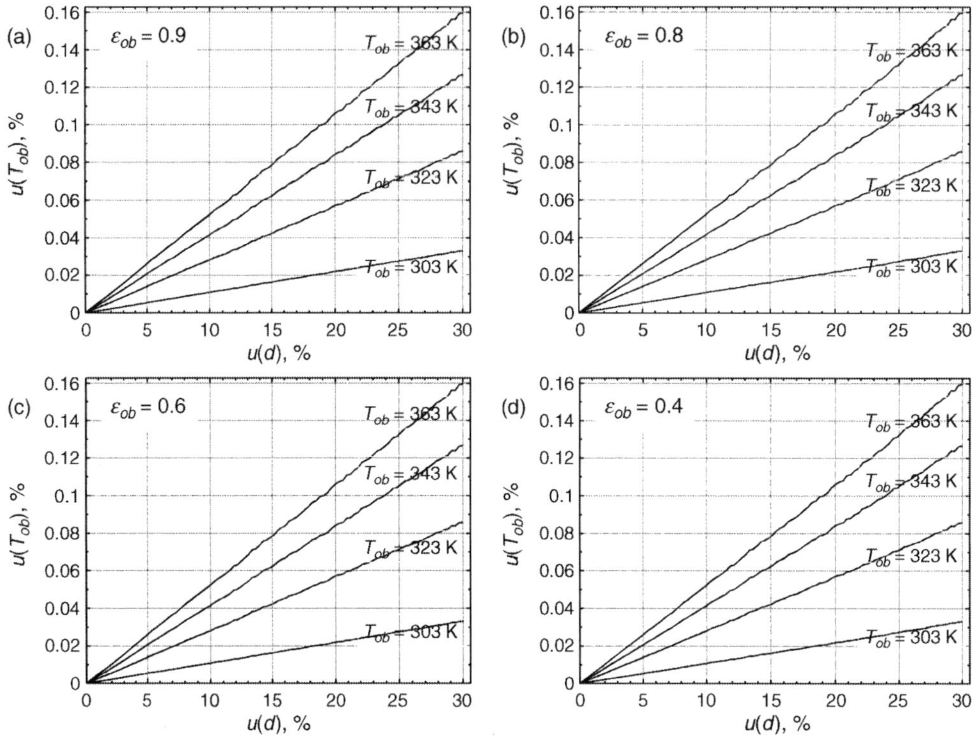

Figure 5.32 Component of the relative standard uncertainty associated with camera-to-object distance d, assuming a uniform distribution. Results for $d = 100$ m

Table 5.4 Estimates of the input variables assumed in analysis of influence of correlations among the inputs of models (3.17) and (3.15) (Figures 3.10b–d) on the relative combined standard uncertainty $u_c(T_{ob})$, %

Object emissivity (ε_{ob})	Ambient temperature (T_o), K	Atmospheric temperature (T_{atm}), K	Relative humidity (ω)	Camera-to-object distance (d), m
0.9, 0.8, 0.6, 0.4	293	293	0.5	50, 100

Table 5.5 Relative standard uncertainties of the input variables assumed in analysis of influence of correlations among the inputs of models (3.17) and (3.15) (Figures 3.10b–d) on the relative combined standard uncertainty $u_c(T_{ob})$, %

Object emissivity (ε_{ob})	Ambient temperature (T_o)	Atmospheric temperature (T_{atm})	Relative humidity (ω)	Camera-to-object distance (d)
10%	10%	10%	10%	10%

interesting cases of measurement conditions are discussed in Minkina and Dudzik (2005), Dudzik (2005) and Dudzik and Minkina (2007).

5.4.2 *Correlations Among Individual Input Variables of Infrared Camera Model and Atmospheric Transmission Model*

Simulations of the combined standard uncertainty $u_c(T_{ob})$ of the object temperature, assuming correlations between pairs of the random variables of models (3.17) and (3.15) (Figure 3.10), are presented in Figures 5.33–5.52 as graphs of $u_c(T_{ob})$ versus correlation coefficient ρ for two specified input variables.

Figures 5.33 and 5.34 show the results of correlation between the variables representing object emissivity ε_{ob} and ambient temperature T_o for two camera-to-object distances: $d = 50$ and 100 m. The values of d assumed in the simulations follow from practical circumstances: on the one hand, for a short camera-to-object distance, the influence of the atmospheric transmission can be neglected; on the other hand, distance $d \approx 100$ m seems to be an upper limit for most typical inspections in infrared thermography.

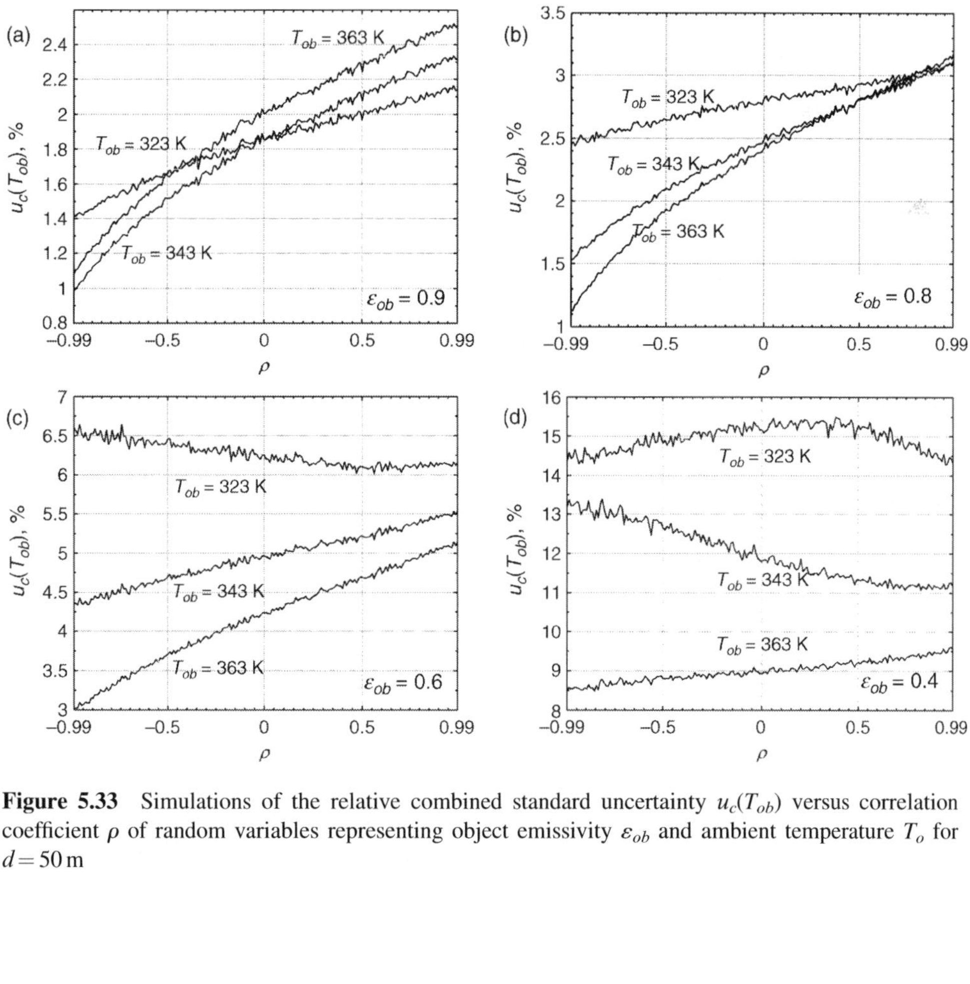

Figure 5.33 Simulations of the relative combined standard uncertainty $u_c(T_{ob})$ versus correlation coefficient ρ of random variables representing object emissivity ε_{ob} and ambient temperature T_o for $d = 50$ m

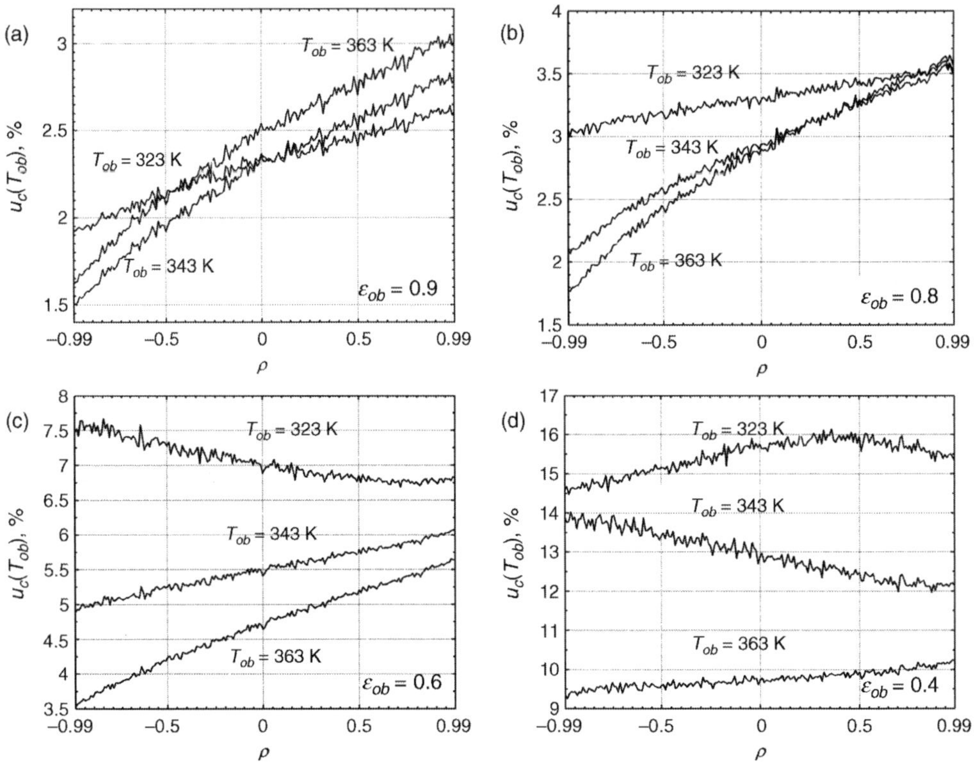

Figure 5.34 Simulations of the relative combined standard uncertainty $u_c(T_{ob})$ versus correlation coefficient ρ of random variables representing object emissivity ε_{ob} and ambient temperature T_o for $d = 100$ m

Simulations of combined standard uncertainty $u_c(T_{ob})$ versus correlation coefficient ρ of random variables representing object emissivity ε_{ob} and atmospheric temperature T_{atm}, for $d = 50$ and $d = 100$ m, are presented in Figures 5.35 and 5.36 respectively.

Simulations of combined standard uncertainty $u_c(T_{ob})$ versus correlation coefficient ρ, assuming correlation between random variables representing object emissivity ε_{ob} and atmospheric relative humidity ω, for two camera-to-object distances of $d = 50$ and 100 m, are shown in Figures 5.37 and 5.38.

Figures 5.39 and 5.40 show simulations of combined standard uncertainty $u_c(T_{ob})$ versus correlation coefficient ρ, assuming correlation between random variables representing object emissivity ε_{ob} and camera-to-object distance d for $d = 50$ and 100 m respectively.

Simulations of combined standard uncertainty $u_c(T_{ob})$ versus correlation coefficient ρ, assuming correlation between random variables representing ambient temperature T_o and atmospheric temperature T_{atm}, for two distances $d = 50$ and 100 m, are shown in Figures 5.41 and 5.42.

Figures 5.43 and 5.44 show simulations of combined standard uncertainty $u_c(T_{ob})$ versus correlation coefficient ρ, assuming correlation between random variables representing

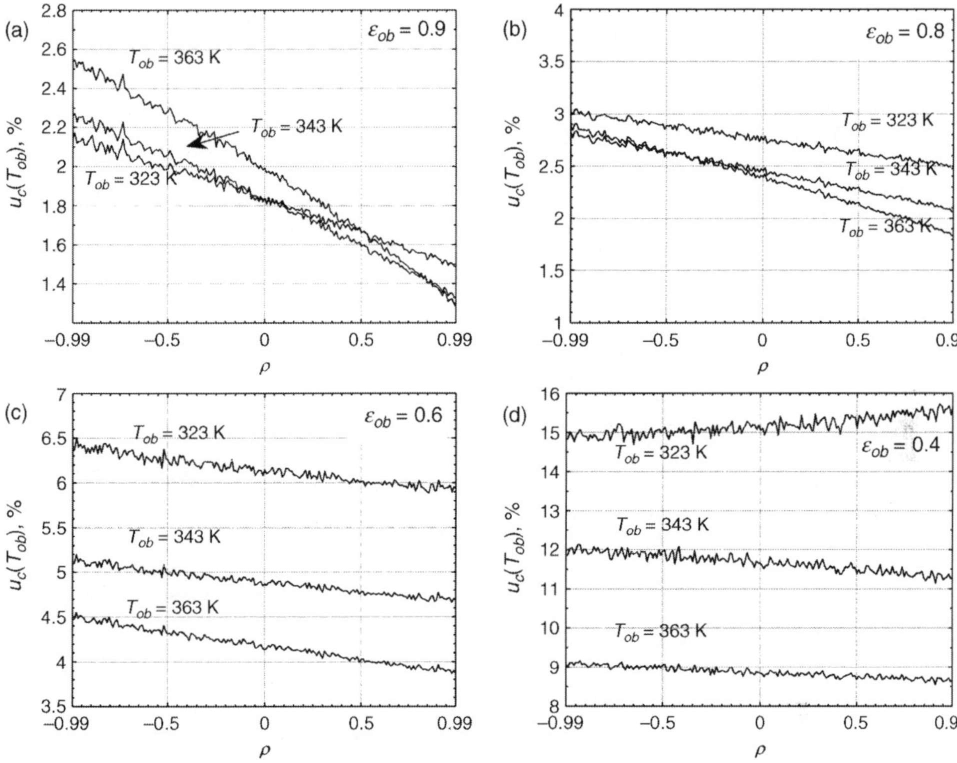

Figure 5.35 Simulations of the relative combined standard uncertainty $u_c(T_{ob})$ versus correlation coefficient ρ of random variables representing object emissivity ε_{ob} and atmospheric temperature T_{atm} for $d = 50$ m

ambient temperature T_o and atmosphere relative humidity ω, for two distances $d = 50$ and 100 m respectively.

Simulations of combined standard uncertainty $u_c(T_{ob})$ versus correlation coefficient ρ, assuming correlation between random variables representing ambient temperature T_o and camera-to-object distance d, for $d = 50$ and 100 m, are shown in Figures 5.45 and 5.46.

Figures 5.47 and 5.48 show simulations of combined standard uncertainty $u_c(T_{ob})$ versus correlation coefficient ρ, assuming correlation between random variables representing atmospheric temperature T_{atm} and atmospheric relative humidity ω, for two distances $d = 50$ and 100 m respectively.

Simulations of combined standard uncertainty $u_c(T_{ob})$ versus correlation coefficient ρ, assuming correlation between random variables representing atmospheric temperature T_{atm} and camera-to-object distance d, for $d = 50$ and 100 m, are shown in Figures 5.49 and 5.50.

Figures 5.51 and 5.52 show simulations of combined standard uncertainty $u_c(T_{ob})$ versus correlation coefficient ρ, assuming correlation between random variables representing atmospheric relative humidity ω and camera-to-object distance d, for $d = 50$ and 100 m respectively.

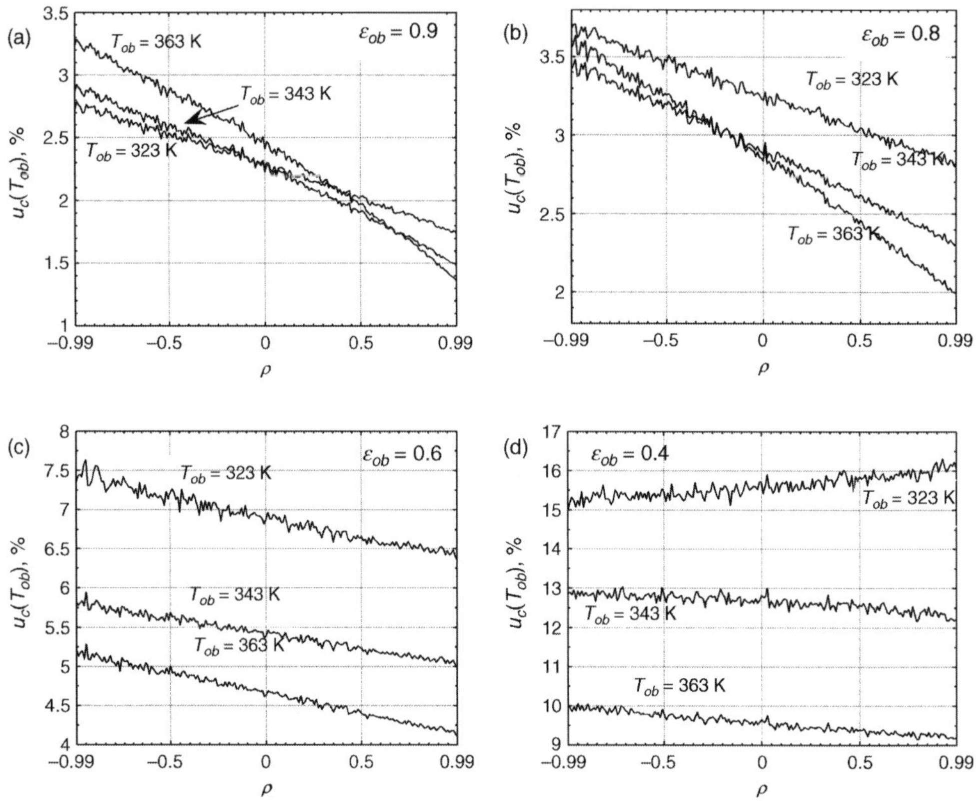

Figure 5.36 Simulations of the relative combined standard uncertainty $u_c(T_{ob})$ versus correlation coefficient ρ of random variables representing object emissivity ε_{ob} and atmospheric temperature T_{atm} for $d = 100$ m

5.4.3 Conclusions

Analysis of the presented simulations of combined standard uncertainty $u_c(T_{ob})$, taking into consideration possible cross-correlations between the input quantities of the infrared camera measurement model, allows us to draw the following conclusions:

- The relative combined standard uncertainty $u_c(T_{ob})$ of model (3.17) with model (3.15) (Figures 3.10b–d) depends mostly on the correlation between the input variables representing object emissivity ε_{ob} and ambient temperature T_o – Figures 5.33 and 5.34.
- The relationship between combined uncertainty $u_c(T_{ob})$ and correlation between ε_{ob} and T_o is dependent on the camera-to-object distance. Comparing, for example, Figures Figure 5.33a and Figure 5.34a, we can see that the graphs of $u_c(T_{ob})$, for $d = 50$ and 100 m, are different.
- The influence of correlation between ε_{ob} and T_o on uncertainty $u_c(T_{ob})$ depends on the object temperature T_{ob} – Figures 5.33 and 5.34.
- The influence of correlation between ε_{ob} and T_o on uncertainty $u_c(T_{ob})$ depends on ε_{ob} – compare, for example, Figures 5.33a–d.

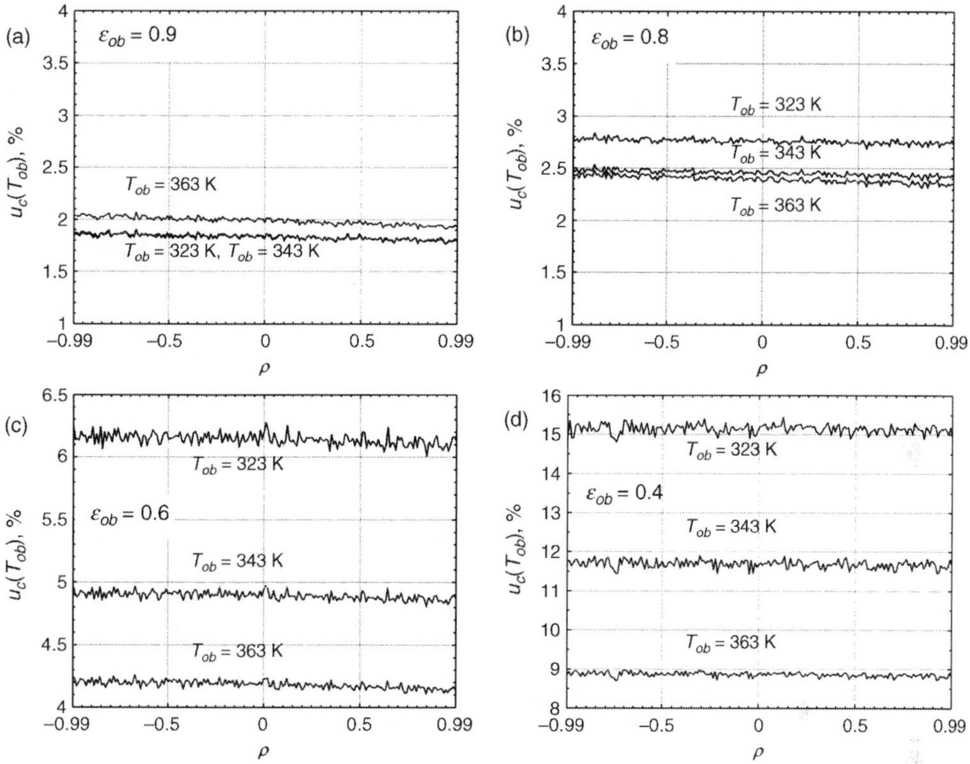

Figure 5.37 Simulations of the relative combined standard uncertainty $u_c(T_{ob})$ versus correlation coefficient ρ of random variables representing object emissivity ε_{ob} and atmospheric relative humidity ω for $d = 50$ m

- The graphs presented in Figures 5.35 and 5.36 show that combined uncertainty $u_c(T_{ob})$ depends on the correlation between the input variables representing object emissivity ε_{ob} and atmospheric temperature T_{atm}. In addition, we can see from Figures 5.35 and 5.36 that the influence of this correlation diminishes with decreasing ε_{ob}.
- Figures 5.35 and 5.36 show that influence of object temperature T_{ob} on the function of combined uncertainty $u_c(T_{ob})$ versus correlation coefficient ρ increases with decreasing ε_{ob}. This tendency can be noticed by comparing, for example, Figures 5.36a and c. The graphs in these figures diverge with a decrease in object emissivity ε_{ob}. Such an effect can be observed for all pairs of correlated input variables. Thus, we can conclude that the lower the object emissivity, the more the function $u_c(T_{ob}) = f(\rho)$ depends on the object temperature. This results from the strong dependence of all components of the combined standard uncertainty on object temperature T_{ob}.
- From Figures 5.41 and 5.42 we can observe that combined uncertainty $u_c(T_{ob})$ depends also on the correlation between ambient temperature T_o and atmospheric temperature T_{atm}. A comparison of, for example, Figures 5.41a–d shows that the influence of correlation between T_o and T_{atm} on uncertainty $u_c(T_{ob})$ decreases with decreasing ε_{ob}. From Figures 5.41 and 5.42 we can also see that camera-to-object distance d, in principle, does not influence

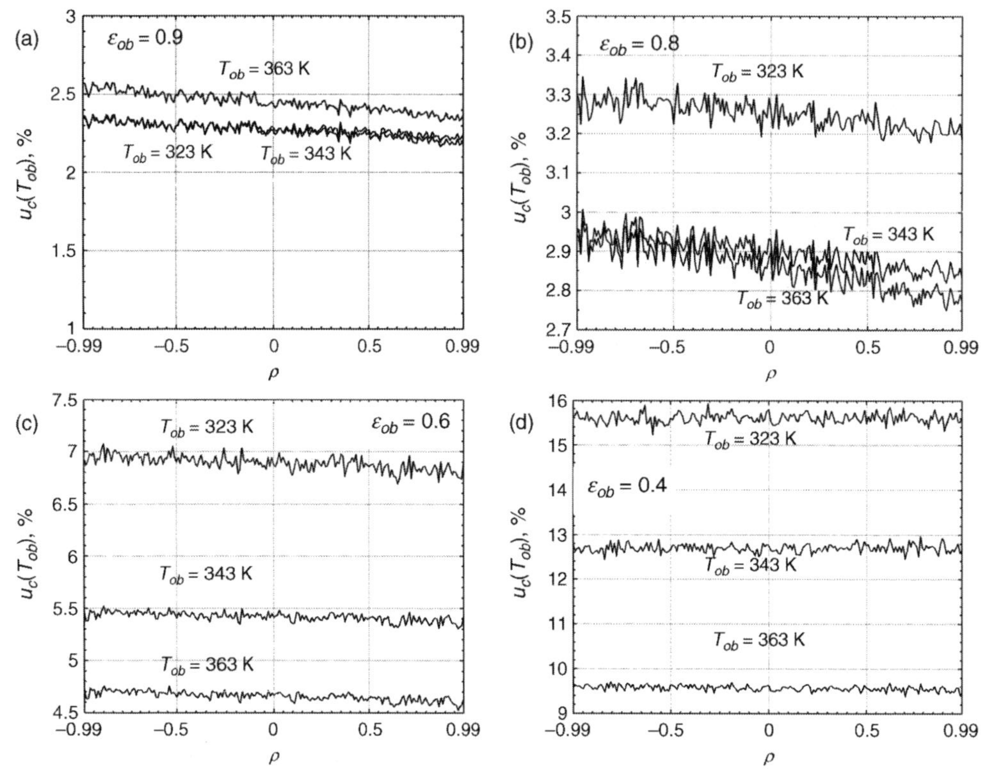

Figure 5.38 Simulations of the relative combined standard uncertainty $u_c(T_{ob})$ versus correlation coefficient ρ of random variables representing object emissivity ε_{ob} and atmospheric relative humidity ω for $d = 100$ m

the graphs of the combined uncertainty. The values of d assumed in simulations follow from practical circumstances: for a short camera-to-object distance, the influence of the atmospheric transmission can be neglected. On the other hand, distance $d \approx 100$ m seems to be an upper limit for most inspections by infrared thermography.

- Analysis of the presented simulation results leads to the observation that combined uncertainty $u_c(T_{ob})$ depends essentially on only the correlations between the pairs of input variables mentioned above: that is, ε_{ob} with T_o, ε_{ob} with T_{atm} and T_o with T_{atm}. All other cases of cross-correlations between two input variables practically do not affect the combined standard uncertainty – see Figures 5.37–5.40.

Summing up the above considerations on the influence of correlations among the input variables of models (3.17) and (3.15) (Figures 3.10b–d) on the relative uncertainty $u_c(T_{ob})$ of temperature measurement by means of infrared thermography, we can state that this influence depends strongly on the measurement conditions. Looking at the graphs of $u_c(T_{ob})$, we can conclude that the values obtained do not contribute much to the total uncertainty budget. However, the results concern the so-called 1-sigma uncertainty determined at a relatively low confidence level. In order to determine the expanded uncertainty, the standard uncertainty

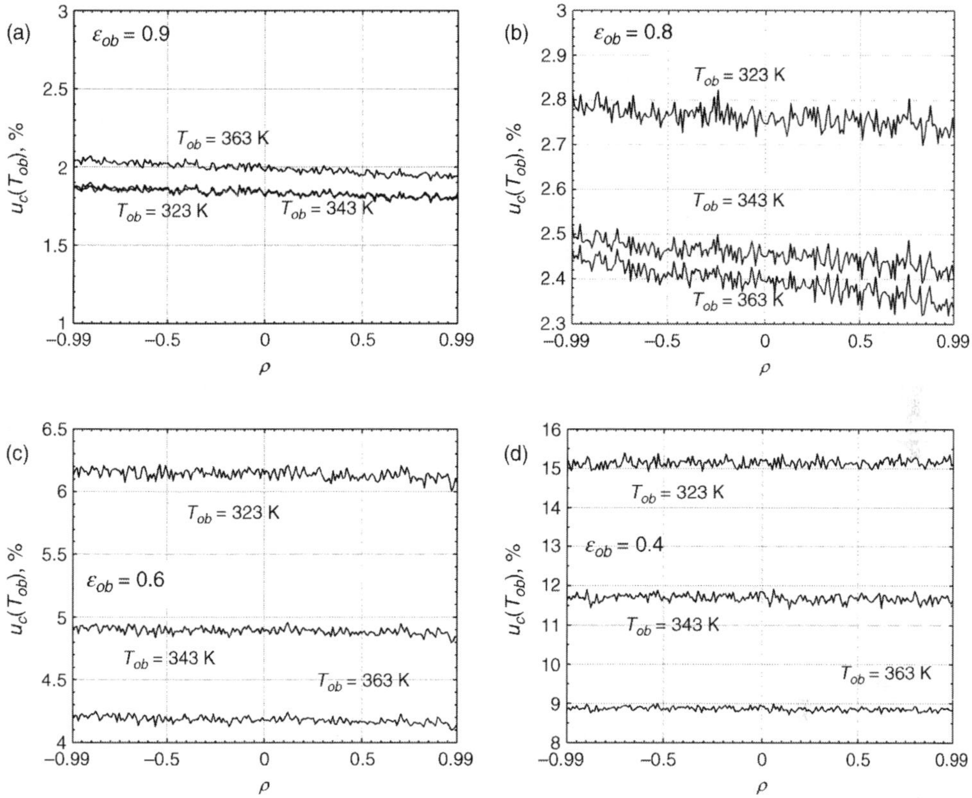

Figure 5.39 Simulations of the relative combined standard uncertainty $u_c(T_{ob})$ versus correlation coefficient ρ of random variables representing object emissivity ε_{ob} and camera-to-object distance d, for $d = 50$ m

should be multiplied by an expansion factor – see (1.20) in Section 1.2. This results in an increase in the influence of correlations on the uncertainty. In addition, our analysis deals with relative uncertainties. The actual influence of correlations can be obtained by recalculating the values shown in the graphs on absolute uncertainties (expressed in kelvin, K). For example, the relative uncertainty in Figure 5.33d for $T_{ob} = 343$ K is about 12%, that is about 41 K in absolute units, when the input variables are uncorrelated ($\rho = 0$). From the same graph ($T_{ob} = 343$ K), we can see that for $\rho = 0.99$ (high positive correlation), the relative uncertainty is about 11.5%, so the absolute uncertainty is about 39 K. For $\rho = -0.99$ (high negative correlation), the relative uncertainty is about 13.5%, so the absolute uncertainty is about 46 K. This example shows how large the divergences between absolute values of the combined standard uncertainty are. The reason for focusing on the dependence of combined uncertainty $u_c(T_{ob})$ on correlation between emissivity ε_{ob} and T_o, and the choice of this pair for the above example, is not purely theoretical. As described in Section 2.3, the emissivity of an object depends on its temperature. Since practical considerations deal with gray bodies

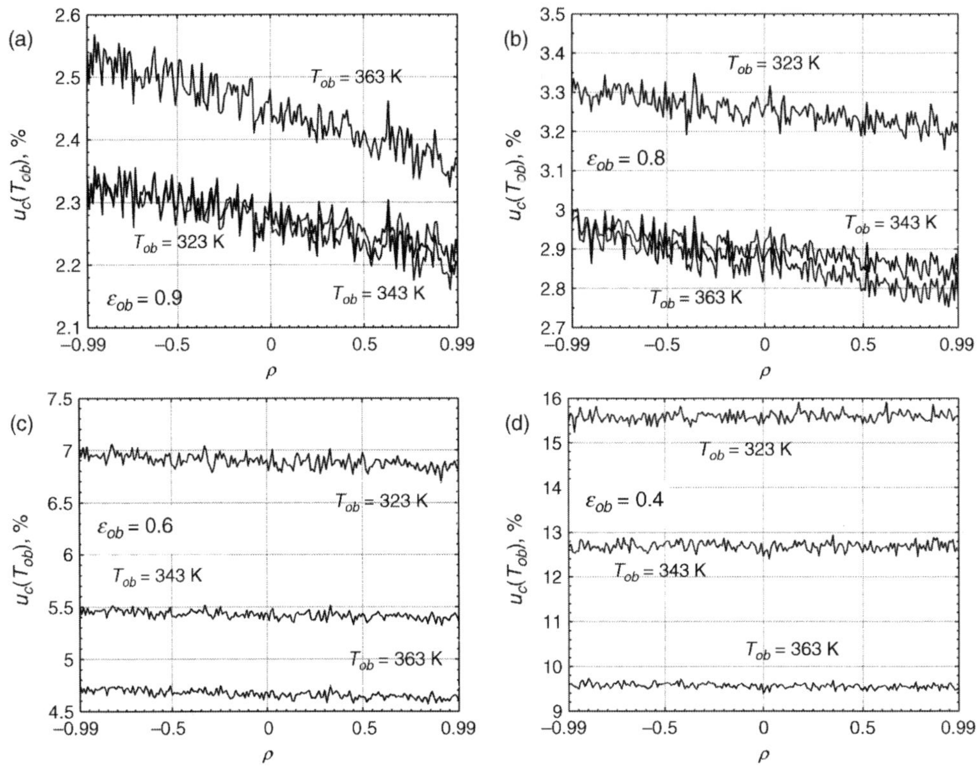

Figure 5.40 Simulations of the relative combined standard uncertainty $u_c(T_{ob})$ versus correlation coefficient ρ of random variables representing object emissivity ε_{ob} and camera-to-object distance d, for $d = 100$ m

(whose $\varepsilon_{ob} < 1$), evaluation of the emissivity depends strongly on the ambient radiation, which in turn depends on T_o. Example 2.3 in Section 2.3 describes the measurement of emissivity in an open measurement chamber. Placing an object in a chamber is intended to make measurement independent of the ambient radiation as much as possible. Unfortunately, complete separation of an object from the ambient is impossible, so the ambient radiation (and consequently ambient temperature T_o) affects infrared thermography measurement to some extent. Taking into consideration the cross-correlation between ε_{ob} and T_o is not a theoretical issue and may be necessary in practical infrared thermography measurements because, as we showed, the influence of this correlation on the combined uncertainty is *significant*. Similar reasoning can be performed for other pairs of input variables if correlations between them are probable. For example, ambient temperature T_o and atmospheric temperature T_{atm} are almost always strongly correlated; T_o is usually assumed to be equal to T_{atm} or strictly dependent on T_{atm}, because the temperature of some or all radiating objects located in the direct neighborhood of an investigated object, that is the ambient temperature, is equal to T_{atm} – Figure 3.12. Hence, one might suspect that the correlation between T_o and T_{atm}

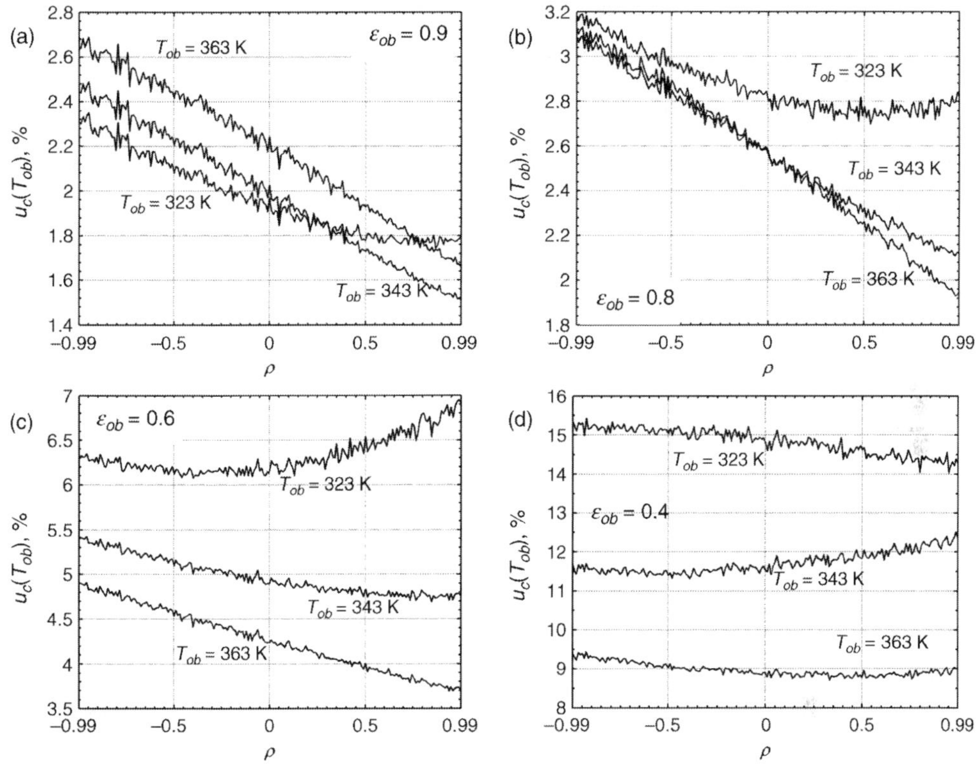

Figure 5.41 Simulations of the relative combined standard uncertainty $u_c(T_{ob})$ versus correlation coefficient ρ of random variables representing ambient temperature T_o and atmospheric temperature T_{atm} for $d = 50$ m

may strongly affect the function of temperature measurement combined uncertainty – see Figures 5.41 and 5.42.

Taking into consideration all possible varieties of measurement conditions in infrared thermography, this chapter could be treated more as an introduction than as an integrated study of the subject. In our opinion, a broad experimental verification of the presented thesis is necessary, apart from the simulation research. This verification should encompass different models of infrared thermography measurements, atmospheric transmission, etc. Only a parallel analysis of simulation and experimental research will lead to the elaboration of a correct model of measurement uncertainty.

The final stage of research of models (3.17) and (3.15) (Figures 3.10b–d) is the evaluation of the combined standard uncertainty of temperature measurement along with a corresponding coverage interval. Evaluations of uncertainty $u_c(T_{ob})$ with calculations of coverage intervals are presented below.

A practical numerical example will serve to illustrate the considerations on the dependence of relative combined standard uncertainty $u_c(T_{ob})$ on correlations among the input variables of models (3.17) and (3.15).

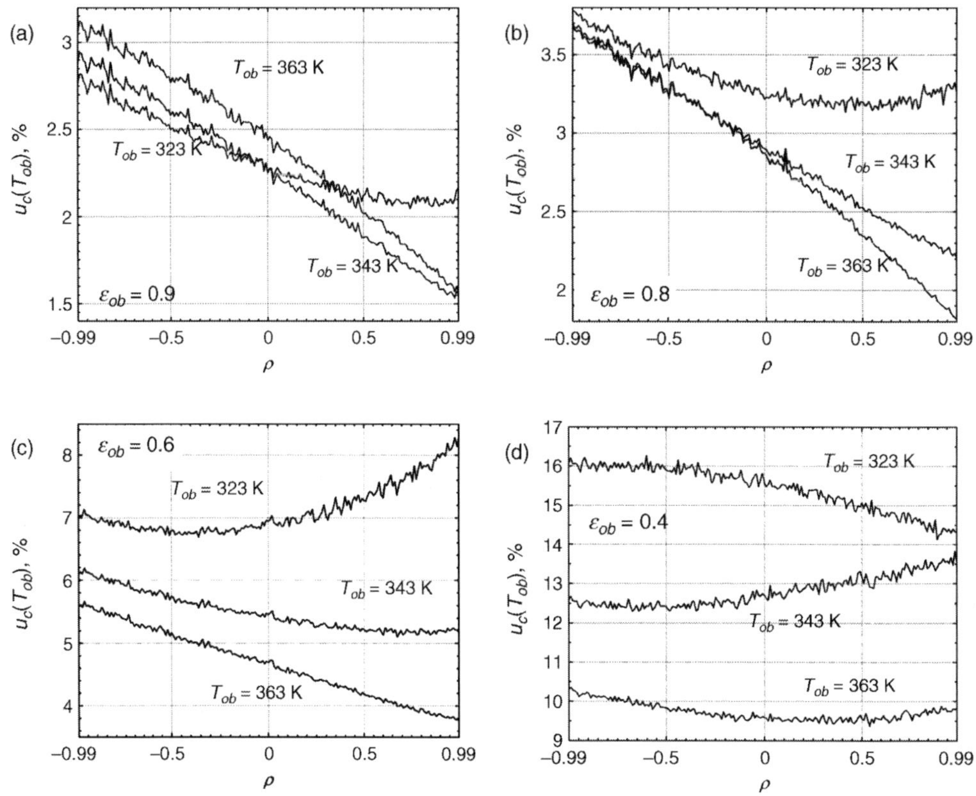

Figure 5.42 Simulations of the relative combined standard uncertainty $u_c(T_{ob})$ versus correlation coefficient ρ of random variables representing ambient temperature T_o and atmospheric temperature T_{atm} for $d = 100$ m

Example 5.1 *Evaluation of the relative combined standard uncertainty taking into account correlations between two selected input variables of model (3.17) and model (3.15) (Figures 3.10b–d)*

Uncertainty $u_c(T_{ob})$ will be calculated for the estimates of the input quantities (i.e. measurement conditions) given in Table 5.6, assuming relative standard uncertainties of these quantities as in Table 5.5.

In this example we assume that correlation occurs only between object emissivity ε_{ob} and atmospheric temperature T_{atm}. Let us also assume that in order to establish the estimates of ε_{ob}

Table 5.6 Estimates of the input quantities assumed in Example 5.1

Object emissivity (ε_{ob})	Ambient temperature (T_o), K	Atmospheric temperature (T_{atm}), K	Relative humidity (ω)	Camera-to-object distance (d), m
0.9	293	293	0.5	50

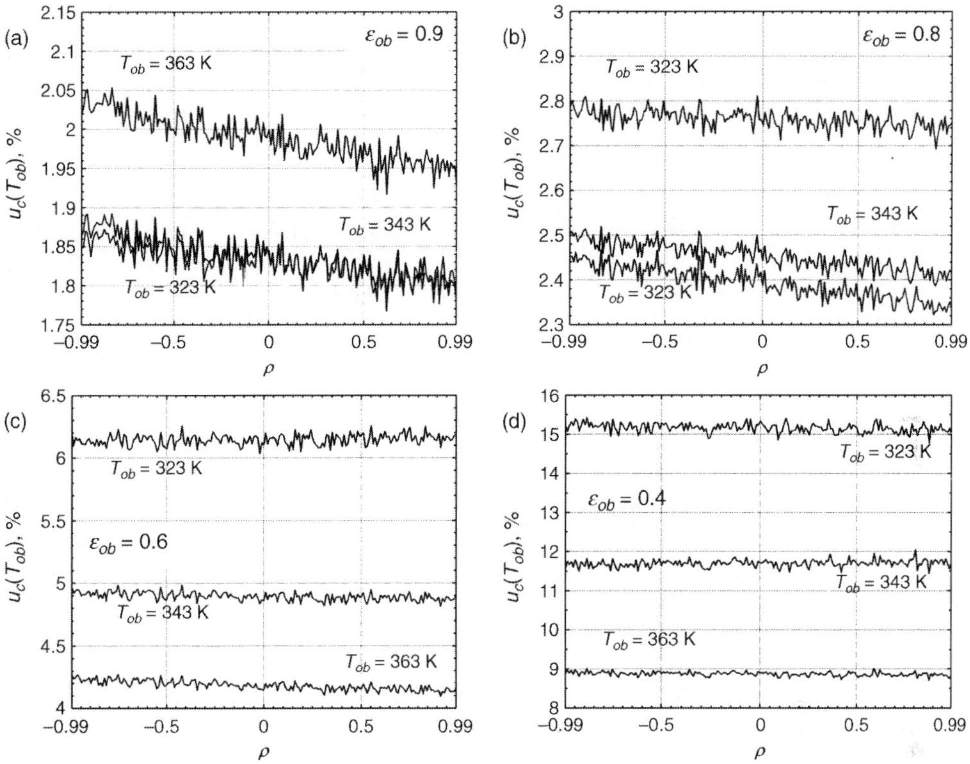

Figure 5.43 Simulations of the relative combined standard uncertainty $u_c(T_{ob})$ versus correlation coefficient ρ of random variables representing ambient temperature T_o and atmospheric relative humidity ω for $d = 50$ m

and T_{atm} to be set in the camera and correlation coefficient ρ of the two variables, a series of 20 measurements of pairs (ε_{ob}, T_{atm}) was performed. The results are shown in Figures 5.53 and 5.54. The horizontal lines indicate mean values (estimates) of the measurements: $\bar{\varepsilon}_{ob} = 0.9$, $\bar{T}_{atm} = 296$ K.

The correlation coefficient is determined using formula (5.12): it yields $\rho = 0.5$. Finally, let us assume that the object temperature read out from the camera was $T_{ob} = 343$ K. Now, from Figure 5.35a, we can see that the relative combined standard uncertainty $u_c(T_{ob}) \approx 1.6\%$.

5.5 Simulations of the Combined Standard Uncertainty for Uncorrelated Input Variables

5.5.1 Introduction

Investigating the accuracy of a measurement model in terms of measurement uncertainty leads to an evaluation of combined standard uncertainty. This combined standard uncer-

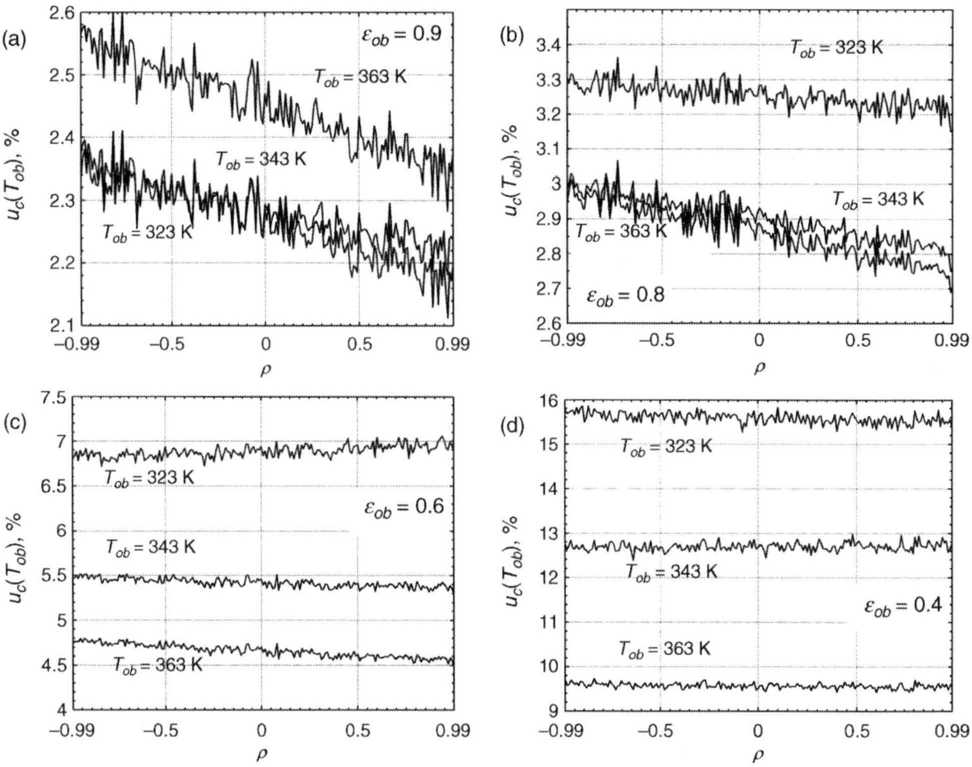

Figure 5.44 Simulations of the relative combined standard uncertainty $u_c(T_{ob})$ versus correlation coefficient ρ of random variables representing ambient temperature T_o and atmospheric relative humidity ω for $d = 100\,\text{m}$

tainty uniquely characterizes measurement accuracy in a statistical sense. As in the case of measurement errors, the combined standard uncertainty is determined from its components associated with individual quantities. Unfortunately, it is usually evaluated at a relatively low confidence level, which means that the probability of finding a measurement result inside the interval determined by the combined uncertainty is relatively low. As mentioned in Section 1.2, in order to increase the probability of finding a measurement result inside a certain interval associated with the uncertainty, we have to know the value of a so-called expansion factor, which extends the uncertainty interval and, consequently, the probability of finding a measurement result inside this extended interval. In Section 1.2 we indicated that the value of the expansion factor depends on the shape of the probability distribution of the measurement model output variable. In general, this probability distribution is unknown, so various methods are applied to approximate the resulting number of degrees of freedom.

In this monograph, the combined standard uncertainty of infrared thermography measurement is evaluated with the method for the propagation of distributions presented in Section 1.3. The components of the combined uncertainty, associated with uncertainties of individual input variables of the temperature measurement model, were discussed in Section 5.3. In

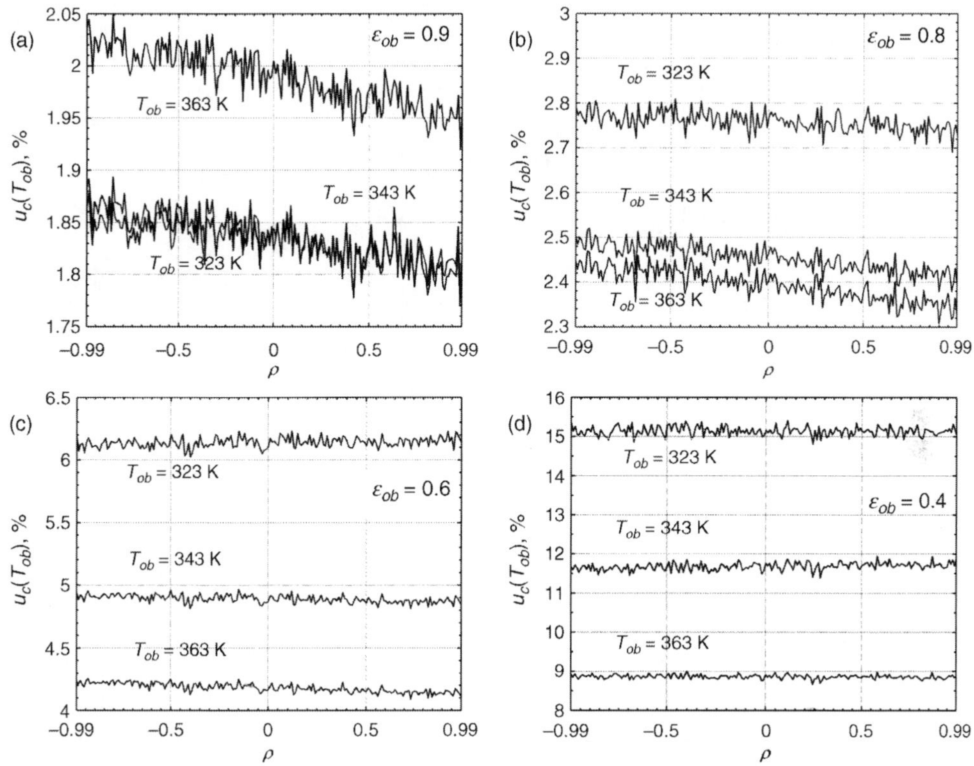

Figure 5.45 Simulations of the relative combined standard uncertainty $u_c(T_{ob})$ versus correlation coefficient ρ of random variables representing ambient temperature T_o and camera-to-object distance d, for $d = 50$ m

Section 5.4 we considered the dependence of the combined uncertainty on correlations among the input variables. In this section we present simulations of combined standard uncertainty $u_c(T_{ob})$ of the object temperature, assuming probability density distributions as described in Section 5.2, namely the logarithmic Gaussian or the uniform distribution. Simulations were carried out for example data of standard uncertainties of the input quantities. The final step in our study of measurement accuracy was the evaluation of the 95% coverage interval. In accordance with the recommendations of Guide (2004), it was determined on the basis of distribution of the output variable of models (3.17) and (3.15) (Figures 3.10b–d), obtained from simulations. The simulation results are presented next.

5.5.2 Simulations of the Combined Standard Uncertainty

The simulations were carried out for 12 cases of different estimates of object emissivity ε_{ob} and its temperature T_{ob} (combinations of four values of the emissivity estimate and three values of the object temperature estimate). As in the investigation of the uncertainty components, simulations were performed for object temperature T_{ob} equal to: 323 K (50 °C), 343 K (70 °C)

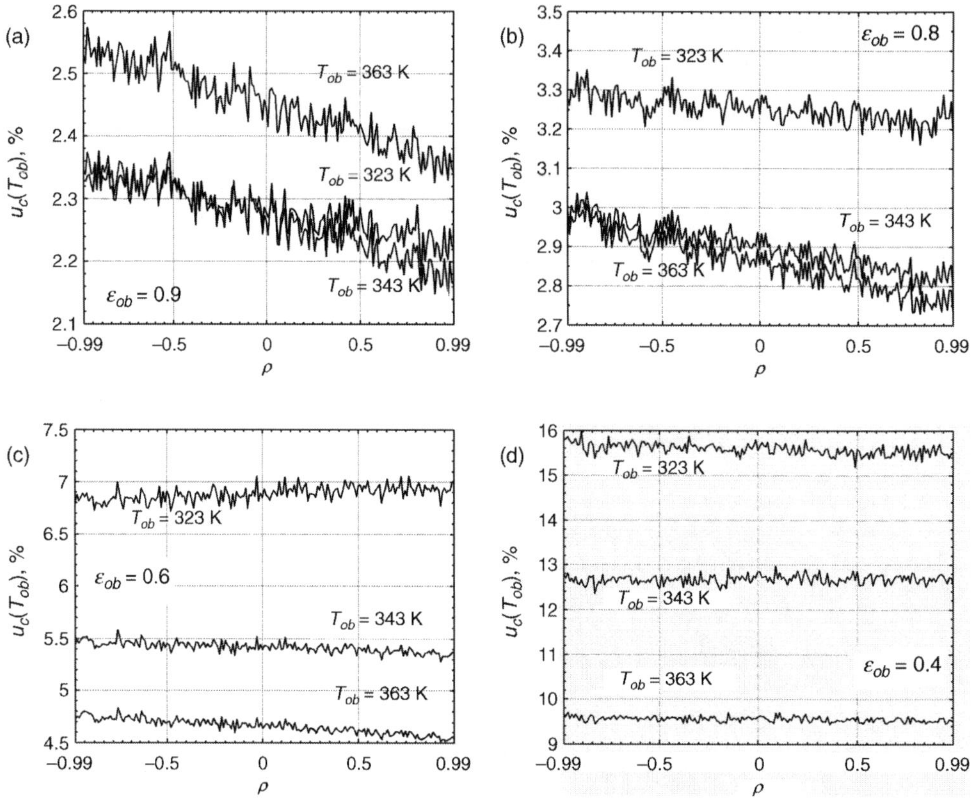

Figure 5.46 Simulations of the relative combined standard uncertainty $u_c(T_{ob})$ versus correlation coefficient ρ of random variables representing ambient temperature T_o and camera-to-object distance d, for $d = 100$ m

and 363 K (90 °C). All these temperatures lie within measurement range I of a typical camera, so the simulation results are valid for this range – other measurement ranges have different calibration constants. For each case the 95% coverage interval $I_{95\%}$ was evaluated. The determined coverage intervals were compared to those calculated for the Gaussian distribution, which is assumed – in most measurements – to be the output variable distribution. The assumption of a Gaussian distribution for the output variables involves expansion coefficient $k = 2$ for the 95% confidence level. In Section 1.3 we mentioned that the width of the 95% coverage interval depends on the symmetry of the output variable probability distribution with respect to its expected value. For example, coefficient α appearing in (1.27) directly determines the quantile order for 95% coverage probability p. For a symmetric distribution $\alpha = 0.025$. To determine how the coverage interval depends on distribution asymmetry, we present, for each case, the 95% coverage interval width as a function of quantile order α (Dudzik and Minkina 2008b). It allows comparison of the minimum width of $I_{95\%}$ to the width for the assumed symmetric distribution of the output variable ($\alpha = 0.025$).

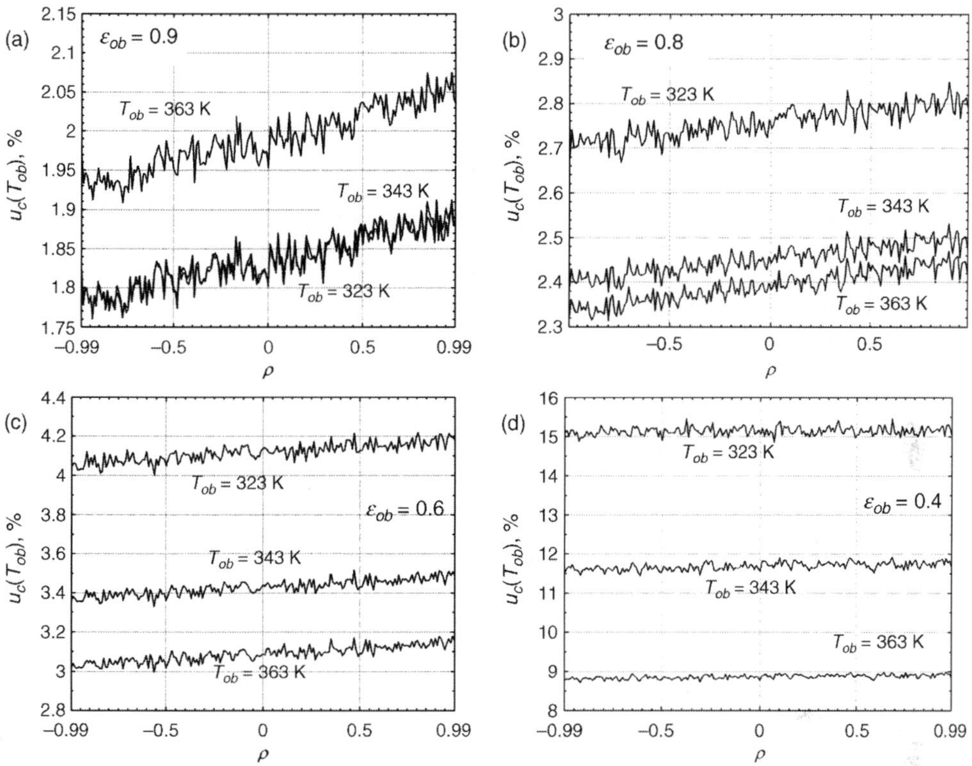

Figure 5.47 Simulations of the relative combined standard uncertainty $u_c(T_{ob})$ versus correlation coefficient ρ of random variables representing atmospheric temperature T_{atm} and relative humidity ω for $d = 50$ m

We carried out simulations of models (3.17) and (3.15) using original software, elaborated in the MATLAB environment. The input data for simulations (i.e. estimates and uncertainties of the input variables) are given in Tables 5.7 and 5.8. For each case we present the normalized histogram of the probability density function $g(T_{ob})$ of the output variable with ends of the 95% coverage interval marked by vertical lines. These limits were determined on the basis of cumulative distributions obtained from simulations and intervals determined assuming the Gaussian probability distribution of the output variable.

Table 5.7 Estimates of the input quantities assumed for analysis of the combined standard uncertainty of models (3.17) and (3.15) (Figures 3.10b–d)

Object emissivity (ε_{ob})	Ambient temperature (T_o)	Atmospheric temperature (T_{atm})	Relative humidity (ω)	Camera-to-object distance (d)
0.9, 0.6, 0.4	293 K	293 K	0.5	10 m

Table 5.8 Standard uncertainties of the input quantity assumed for analysis of the combined standard uncertainty of models (3.17) and (3.15) (Figures 3.10b–d)

Object emissivity (ε_{ob})	Ambient temperature (T_o)	Atmospheric temperature (T_{atm})	Relative humidity (ω)	Camera-to-object distance (d)
0.09, 0.06, 0.04 (10%)	9 K (3%)	9 K (3%)	0.05 (10%)	1 m (10%)

The left-hand side graphs in Figures 5.55-5.72 show the probability density functions of the output variable of models (3.17) and (3.15) with the marked ends of the 95% coverage intervals. These limits were determined from cumulative distributions, assuming the Gaussian distribution for values of ε_{ob} given in Table 5.7 and for $T_{ob} = 323$, 343 and 363 K.

In simulations of the output variable distribution we assumed uniform distributions of the input variables. This follows from the fact that the uniform distribution is the worst case from the point of view of the combined standard uncertainty evaluation. In our research we also considered logarithmic Gaussian distributions of the model input variables. However, for the

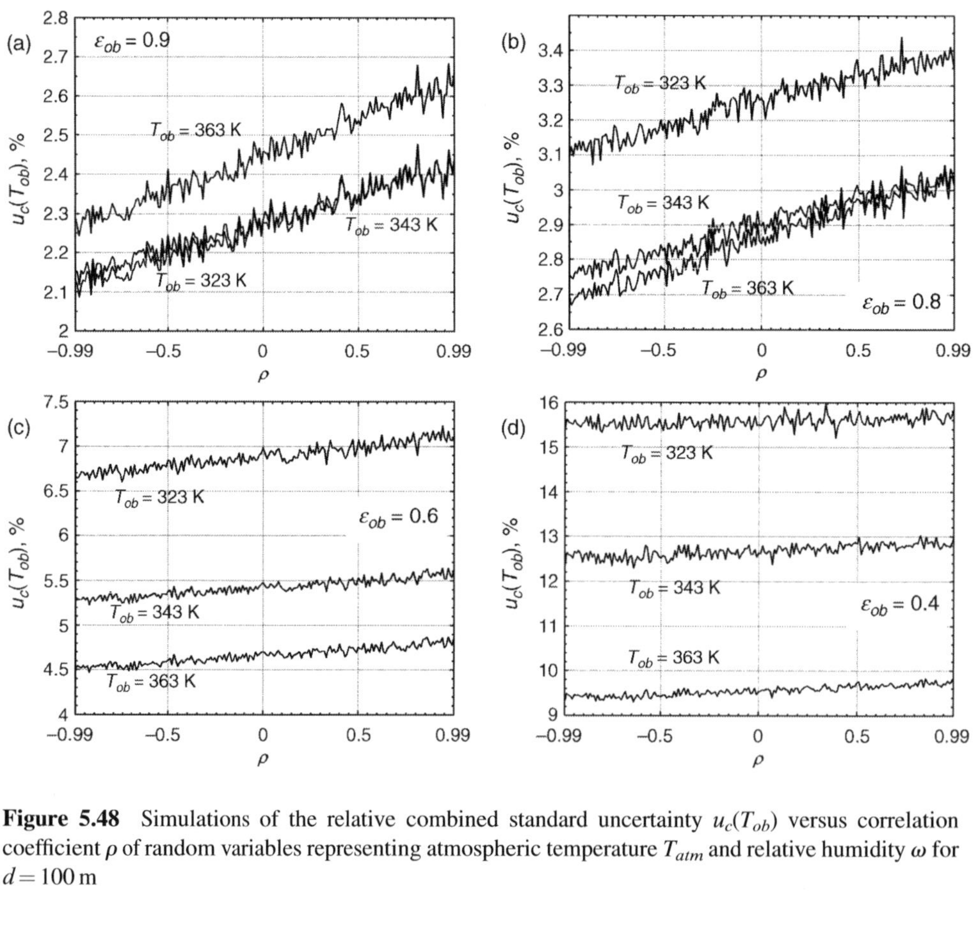

Figure 5.48 Simulations of the relative combined standard uncertainty $u_c(T_{ob})$ versus correlation coefficient ρ of random variables representing atmospheric temperature T_{atm} and relative humidity ω for $d = 100$ m

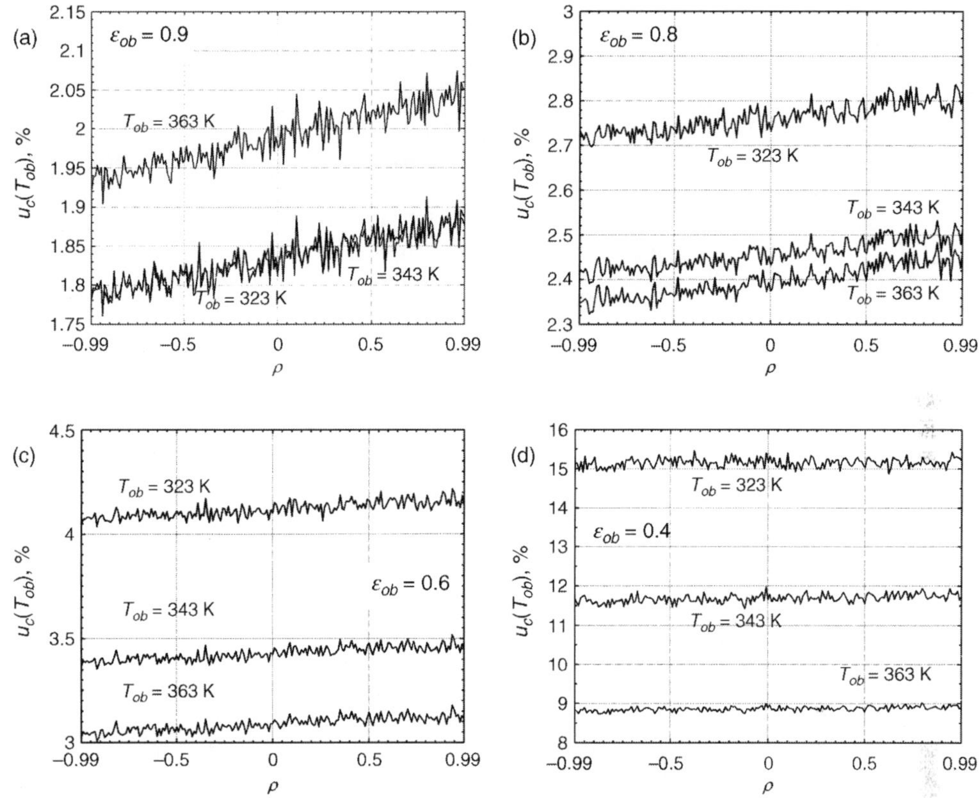

Figure 5.49 Simulations of the relative combined standard uncertainty $u_c(T_{ob})$ versus correlation coefficient ρ of random variables representing atmospheric temperature T_{atm} and camera-to-object distance d, for $d = 50$ m

assumed measurement conditions and simulation input data, we did not notice any significant differences between uncertainties evaluated for one or the other distribution. Nevertheless, we cannot exclude that, for other simulation data, the results depend more on the type of probability distribution assumed for the model input variables.

The dependence of the 95% coverage intervals on quantile order α are presented in the right-hand side graphs in Figures 5.56-5.72. The intervals were determined from cumulative distributions, assuming the Gaussian distribution for values of ε_{ob} given in Table 5.7 and for $T_{ob} = 323$, 343 and 363 K.

The results of simulations of the combined standard uncertainty of infrared thermography measurement are collected in Table 5.9. $I_{95\%\text{-sim}}$ denotes the 95% coverage interval determined for the numerical approximation of the cumulative distribution obtained from simulation. $I_{95\%\text{-norm}}$ denotes the 95% coverage interval determined assuming that the output variable of models (3.17) and (3.15) is subject to the Gaussian distribution.

The values given in parentheses in columns $I_{95\%\text{-sim}}$ and $I_{95\%\text{-norm}}$ of Table 5.9 are the widths of the 95% coverage intervals defined in square brackets in the same row of the table.

Table 5.9 Simulation results for infrared thermography measurement models (3.17) and (3.15), in terms of combined standard uncertainty $u_c(T_{ob})$

ε_{ob}	T_{ob}, K	$u_c(T_{ob})$, K	$I_{95\%\text{-sim}}$	$I_{95\%\text{-norm}}$
0.9	323	2.9 (0.9%)	[319, 329] K (10 K)	[317, 329] K (12 K)
	343	4.3 (1.3%)	[337, 351] K (14 K)	[335, 352] K (17 K)
	363	5.6 (1.5%)	[355, 373] K (18 K)	[352, 375] K (23 K)
0.6	323	5.6 (1.7%)	[313, 333] K (20 K)	[312, 334] K (22 K)
	343	6.0 (1.7%)	[333, 355] K (22 K)	[332, 355] K (23 K)
	363	6.7 (1.8%)	[352, 376] K (24 K)	[350, 377] K (27 K)
0.4	323	11 (3.4%)	[302, 341] K (39 K)	[300, 344] K (44 K)
	343	10 (2.9%)	[324, 361] K (37 K)	[323, 363] K (40 K)
	363	9.9 (2.7%)	[345, 382] K (37 K)	[343, 382] K (39 K)

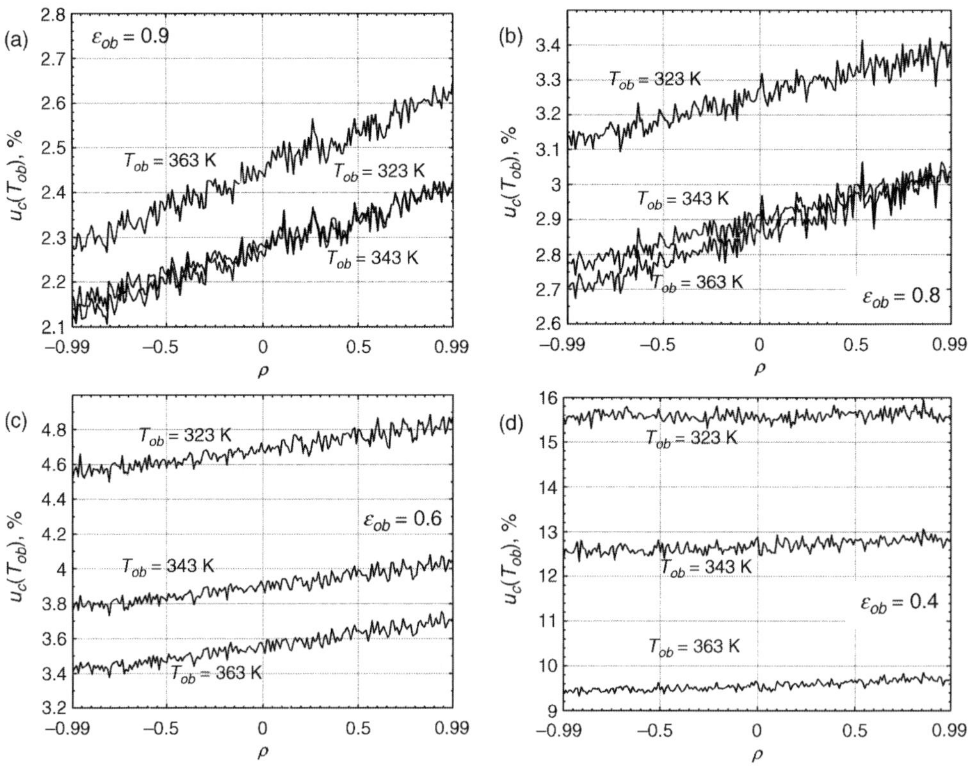

Figure 5.50 Simulations of the relative combined standard uncertainty $u_c(T_{ob})$ versus correlation coefficient ρ of random variables representing atmospheric temperature T_{atm} and camera-to-object distance d, for $d = 100$ m

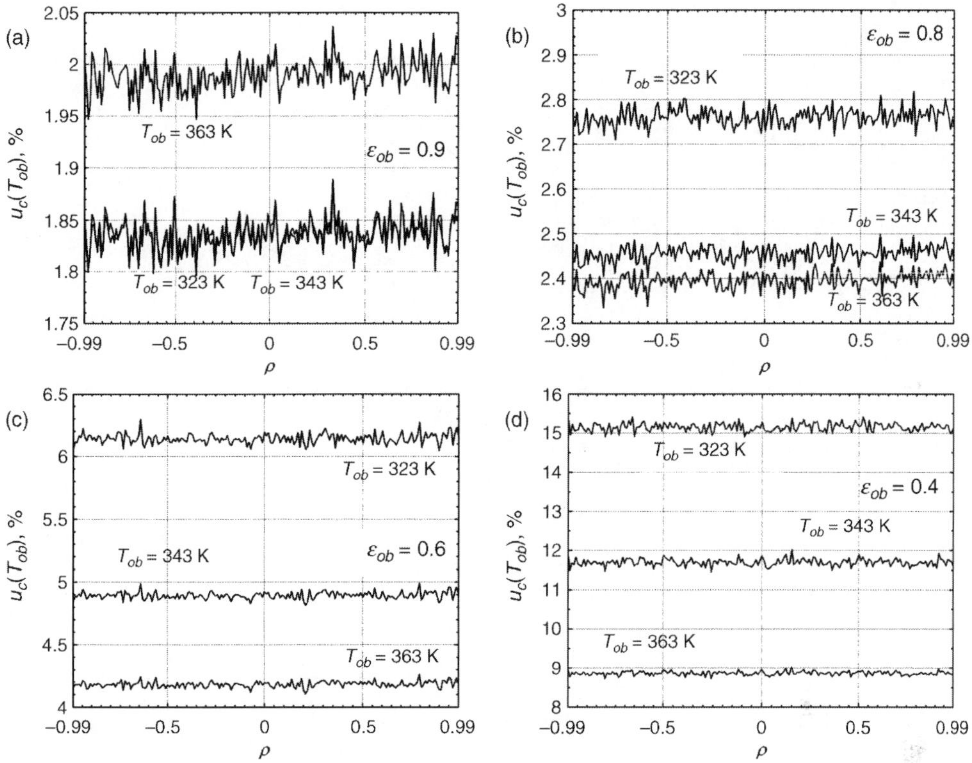

Figure 5.51 Simulations of the relative combined standard uncertainty $u_c(T_{ob})$ versus correlation coefficient ρ of random variables representing atmospheric relative humidity ω and camera-to-object distance d, for $d = 50\,\text{m}$

5.5.3 Conclusions

Analysis of simulations of the combined standard uncertainty $u_c(T_{ob})$ of the object temperature leads to the following conclusions:

- From the results presented in Table 5.9 we can see that for $\varepsilon_{ob} = 0.9$ and 0.6, the combined standard uncertainty increases with increasing object temperature T_{ob}. For $\varepsilon_{ob} = 0.4$ this relationship is the inverse. Additional simulations confirmed that for a low object emissivity, this inversion of tendency is a general phenomenon: for ε_{ob} below 0.5 the combined uncertainty decreases when T_{ob} increases. The results of the additional simulations are not included in this monograph.
- The combined standard uncertainty increases strongly with decreasing object emissivity. From Table 5.9 we can also see that the lower the object temperature, the faster the increase in the uncertainty.

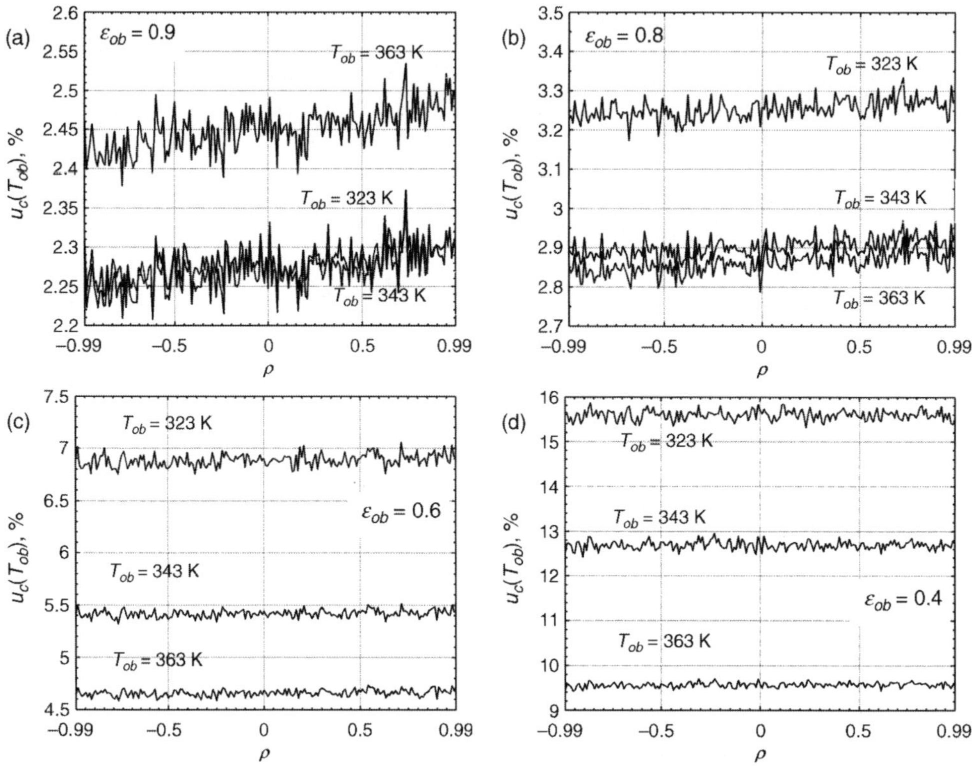

Figure 5.52 Simulations of the relative combined standard uncertainty $u_c(T_{ob})$ versus correlation coefficient ρ of random variables representing atmospheric relative humidity ω and camera-to-object distance d, for $d = 100$ m

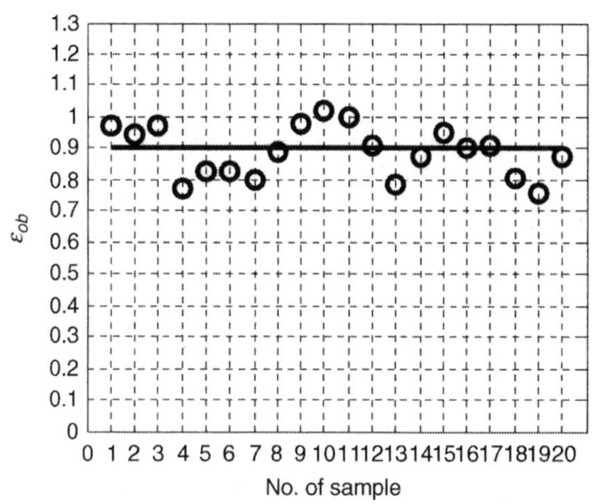

Figure 5.53 Values of 20 measurements of emissivity ε_{ob}

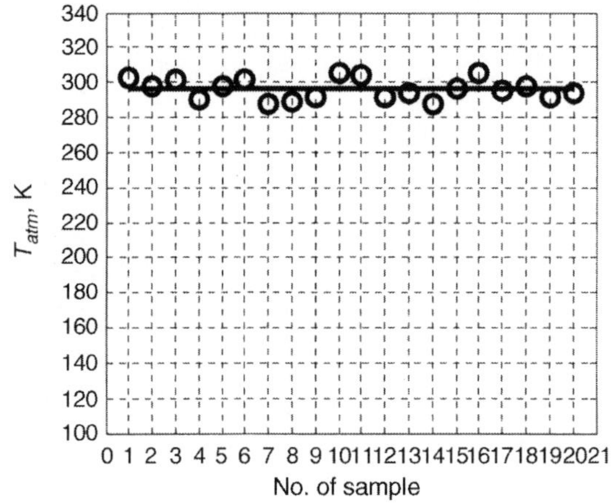

Figure 5.54 Values of 20 measurements of temperature T_{atm}

- The 95% coverage interval is extended strongly with decreasing object emissivity ε_{ob}. For example, Table 5.9 shows that for $\varepsilon_{ob} = 0.9$ and $T_{ob} = 323\,\text{K}$, the width of the interval determined for the cumulative distribution obtained from simulation is 10 K, whereas for $\varepsilon_{ob} = 0.4$ and $T_{ob} = 323\,\text{K}$ it increases to 39 K. This tendency is also visible from a comparison of Figures 5.56 and 5.68.
- Comparing coverage intervals determined on the basis of approximations of the output variable cumulative distributions to coverage intervals determined for Gaussian distributions, we can observe that the differences are insignificant. In general, assuming the Gaussian distribution for the output variable – justified by the central limit theorem – is safe in terms of underestimating the 95% coverage interval. In fact, the dashed lines in Figures 5.55-5.72 (right-hand side graphs), representing the coverage interval widths for the Gaussian

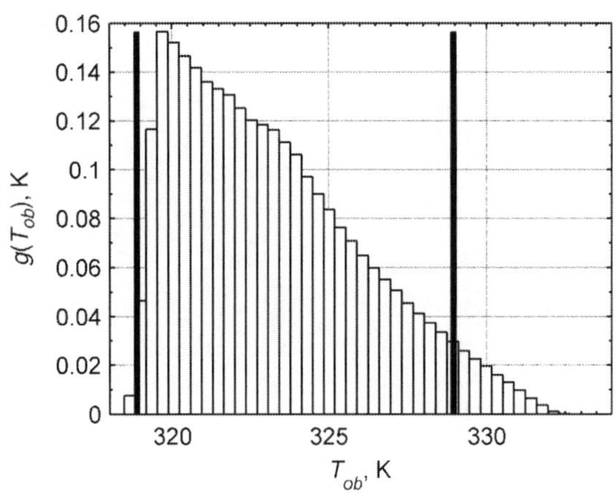

Figure 5.55 Probability density function of output variable T_{ob} of models (3.17) and (3.15), for $T_{ob} = 323\,\text{K}$ and $\varepsilon_{ob} = 0.9$

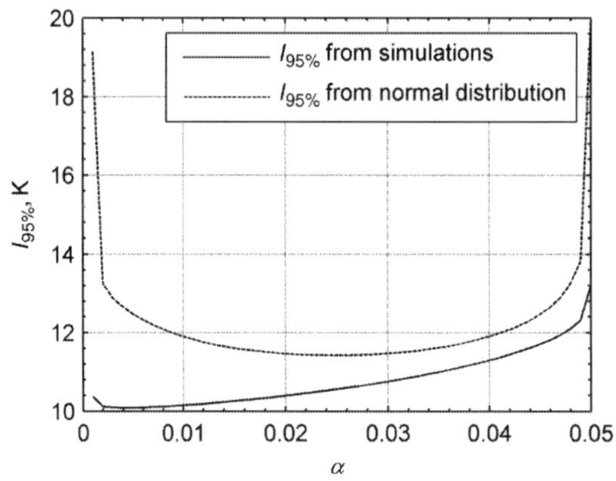

Figure 5.56 The 95% coverage intervals versus quantile order α for $T_{ob} = 323$ K and $\varepsilon_{ob} = 0.9$

distribution, always lie above the solid lines representing the coverage interval widths obtained from simulation.

- Taking into consideration the above observations, we can conclude that for evaluation of the expanded uncertainty of temperature measurement using models (3.17) and (3.15) with the 95% confidence level, it is safe to assume that expansion coefficient $k = 2$, as for the Gaussian distribution. In every case such an assumption will lead to a slight extension of the coverage interval obtained from simulation. In other words, the confidence level of the interval determined for the Gaussian distribution of the output variable of the considered measurement model exceeds slightly 95%, that is the estimation is 'safe'.
- Comparing the graphs corresponding to distributions obtained from simulation to those corresponding to Gaussian distributions, we can see that the 95% coverage interval width is practically independent of the asymmetry of the output variable distribution. Looking at the solid line in Figure 5.68, for $T_{ob} = 323$ K and $\varepsilon_{ob} = 0.4$, we can see that the coverage interval

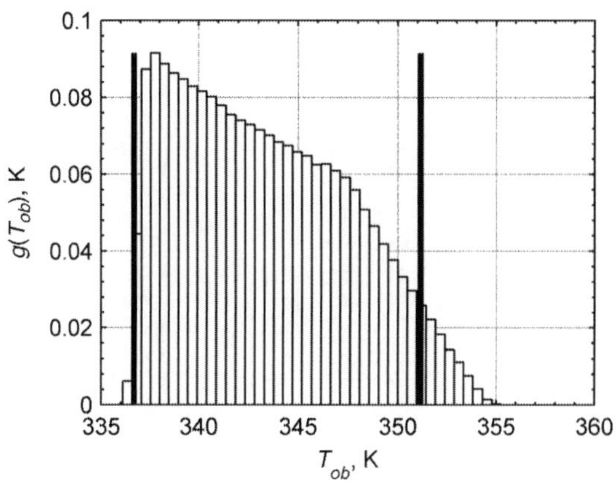

Figure 5.57 Probability density function of output variable T_{ob} of models (3.17) and (3.15), for $T_{ob} = 343$ K and $\varepsilon_{ob} = 0.9$

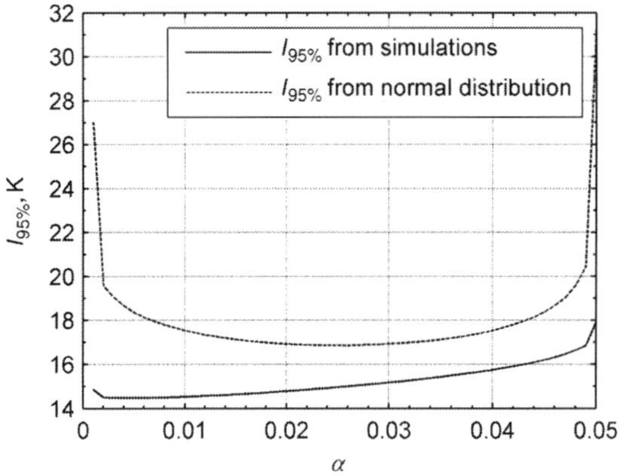

Figure 5.58 The 95% coverage intervals versus quantile order α for $T_{ob} = 343$ K and $\varepsilon_{ob} = 0.9$

is smallest for $\alpha = 0.01$. Comparing the interval widths for $\alpha = 0.01$ and $\alpha = 0.025$ (symmetric distribution) shows that the width changes no more than about 1 K, which is insignificant when we remember that the interval width is above 40 K.

The simulations, whose results are presented here, were conducted to evaluate the combined standard uncertainty of temperature measurement with infrared thermography. The simulations were executed for selected values of the reference quantities and for specified standard uncertainties of the input variables. The standard uncertainties of particular input variables were kept fixed during the simulations, because we did not deal with a sensitivity analysis of models (3.17) and (3.15). The sensitivity analysis was carried out in Section 5.3, where we evaluated components of the combined standard uncertainty associated with individual input variables. In this section we concentrated on evaluating the combined uncertainty in concrete situations of infrared thermography measurements. Analysis of the results leads to an

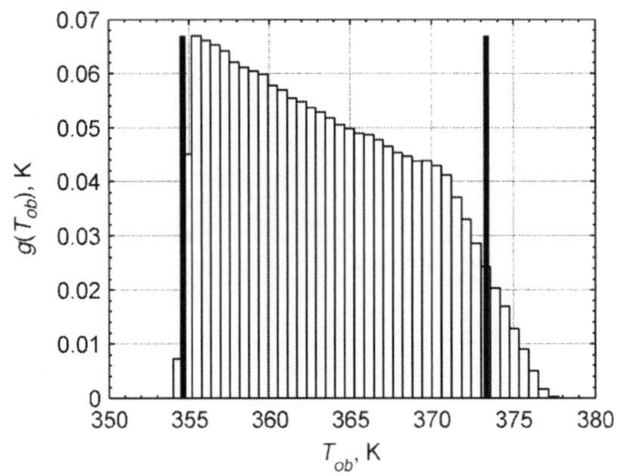

Figure 5.59 Probability density function of output variable T_{ob} of models (3.17) and (3.15), for $T_{ob} = 363$ K and $\varepsilon_{ob} = 0.9$

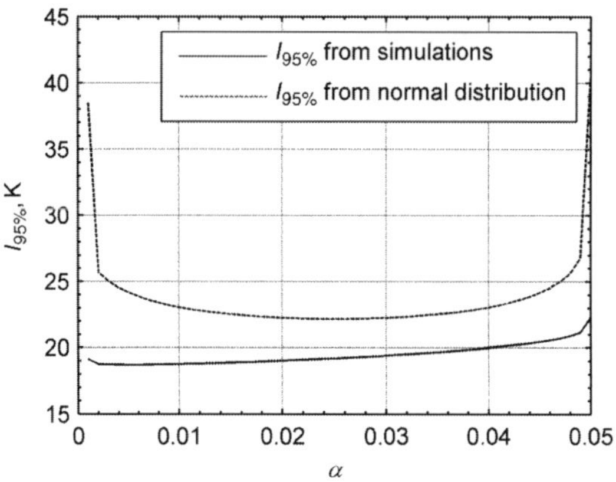

Figure 5.60 The 95% coverage intervals versus quantile order α for $T_{ob} = 363$ K and $\varepsilon_{ob} = 0.9$

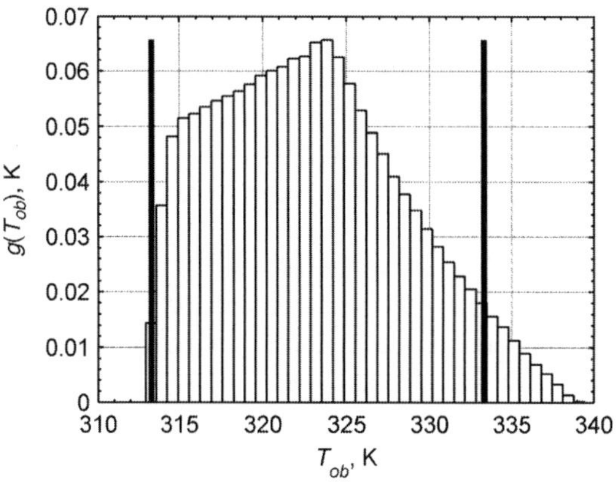

Figure 5.61 Probability density function of output variable T_{ob} of models (3.17) and (3.15), for $T_{ob} = 323$ K and $\varepsilon_{ob} = 0.6$

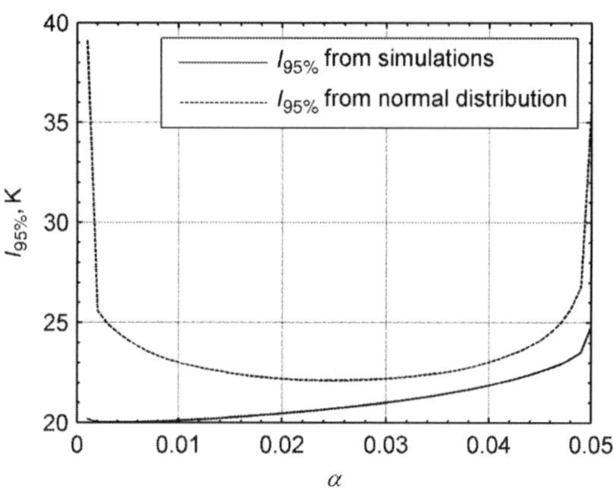

Figure 5.62 The 95% coverage intervals versus quantile order α for $T_{ob} = 323$ K and $\varepsilon_{ob} = 0.6$

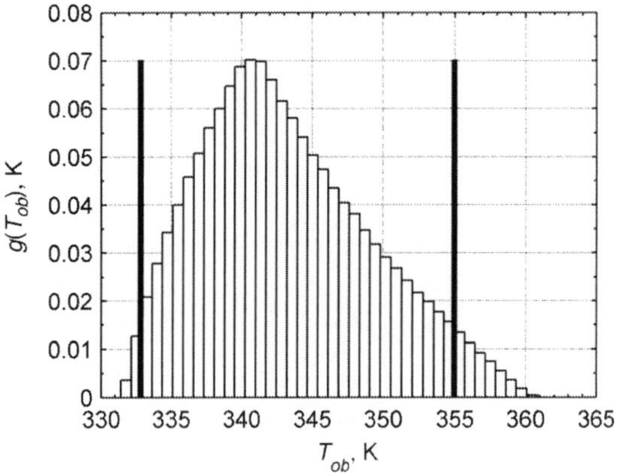

Figure 5.63 Probability density function of output variable T_{ob} of models (3.17) and (3.15), for $T_{ob} = 343$ K and $\varepsilon_{ob} = 0.6$

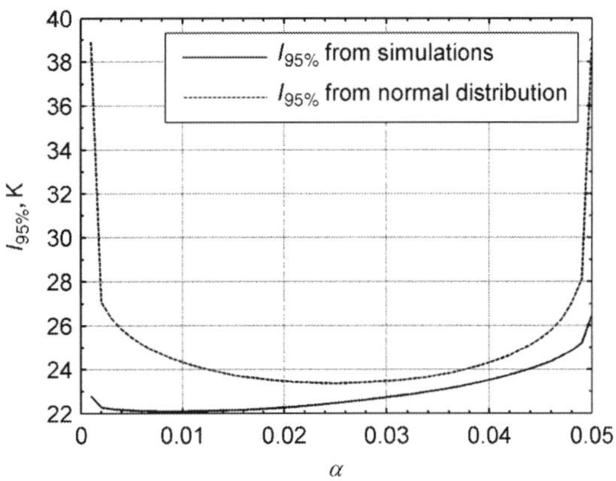

Figure 5.64 The 95% coverage intervals versus quantile order α for $T_{ob} = 343$ K and $\varepsilon_{ob} = 0.6$

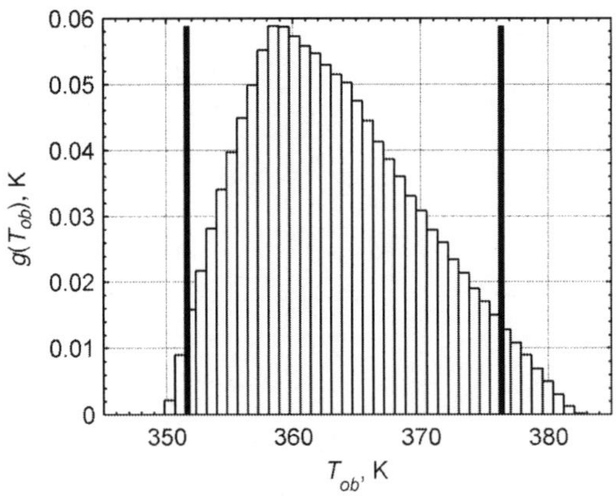

Figure 5.65 Probability density function of output variable T_{ob} of models (3.17) and (3.15), for $T_{ob} = 363$ K and $\varepsilon_{ob} = 0.6$

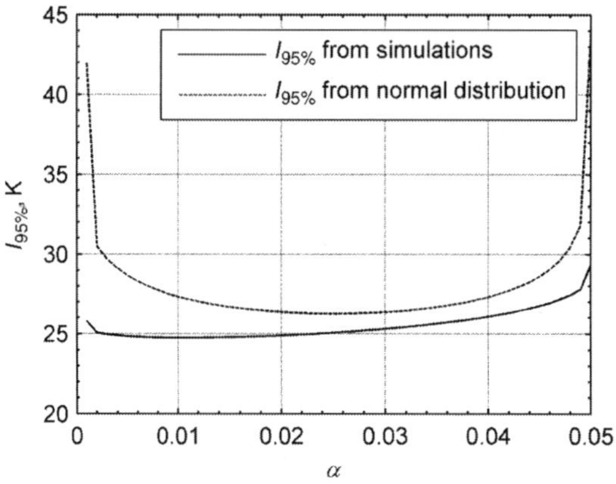

Figure 5.66 The 95% coverage intervals versus quantile order α for $T_{ob} = 363$ K and $\varepsilon_{ob} = 0.6$

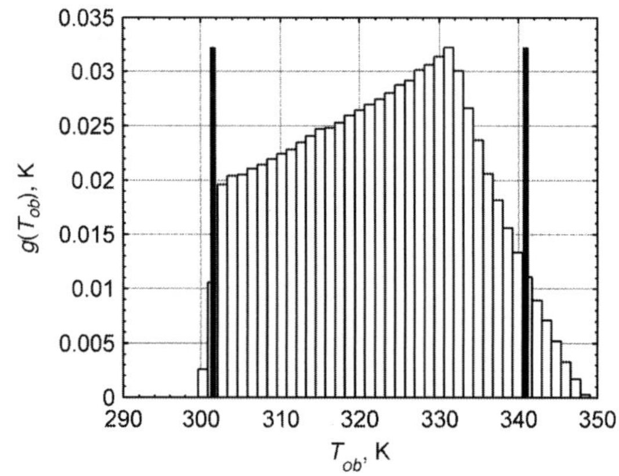

Figure 5.67 Probability density function of output variable T_{ob} of models (3.17) and (3.15), for $T_{ob} = 323$ K and $\varepsilon_{ob} = 0.4$

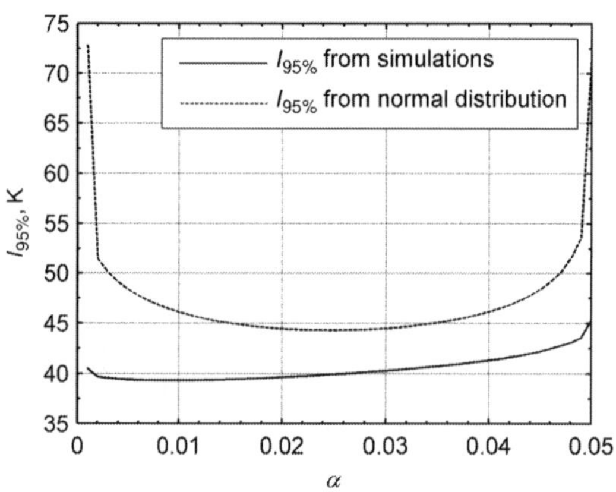

Figure 5.68 The 95% coverage intervals versus quantile order α for $T_{ob} = 323$ K and $\varepsilon_{ob} = 0.4$

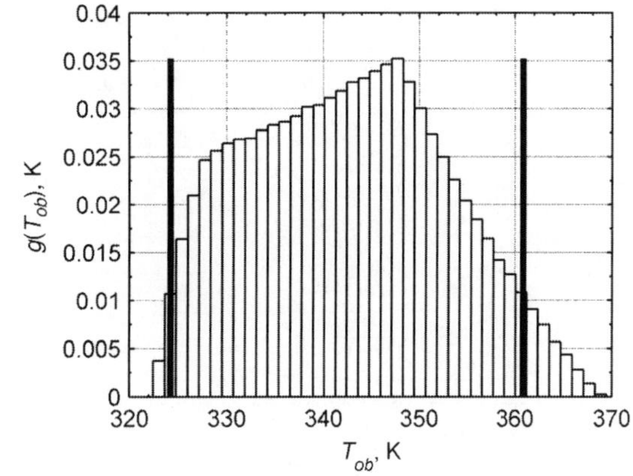

Figure 5.69 Probability density function of output variable T_{ob} of models (3.17) and (3.15), for $T_{ob} = 343$ K and $\varepsilon_{ob} = 0.4$

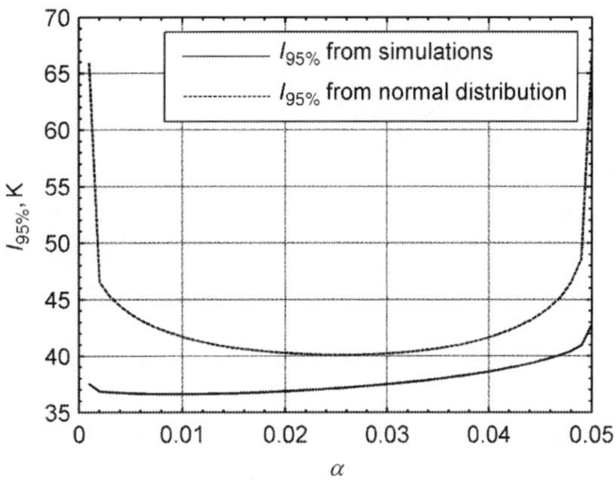

Figure 5.70 The 95% coverage intervals versus quantile order α for $T_{ob} = 343$ K and $\varepsilon_{ob} = 0.4$

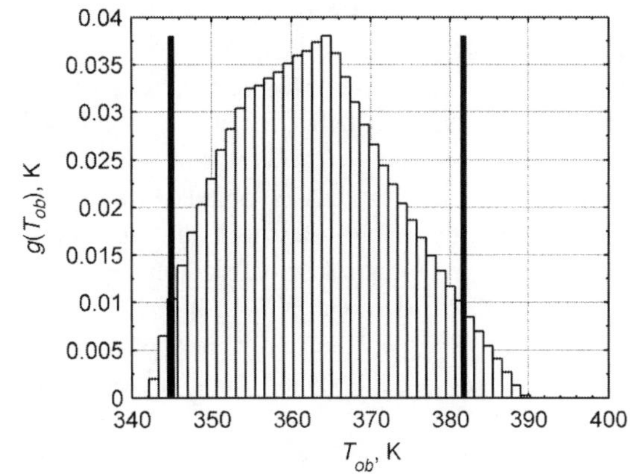

Figure 5.71 Probability density function of output variable T_{ob} of models (3.17) and (3.15), for $T_{ob} = 363$ K and $\varepsilon_{ob} = 0.4$

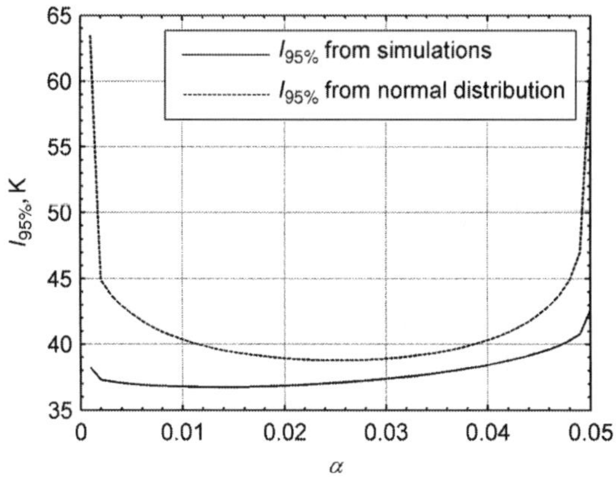

Figure 5.72 The 95% coverage intervals versus quantile order α for $T_{ob} = 363$ K and $\varepsilon_{ob} = 0.4$

interesting conclusion: the combined standard uncertainty of measurement depends on object temperature T_{ob}. For medium and high object emissivity, $\varepsilon_{ob} > 0.5$, the uncertainty increases with T_{ob}, whereas for smaller emissivity, $\varepsilon_{ob} < 0.5$, the tendency is the inverse – the uncertainty decreases with increasing T_{ob}. Consequently, for $\varepsilon_{ob} \approx 0.5$ we should observe that the combined standard uncertainty is independent of the object temperature. In fact, this was confirmed by a simulation investigation of the infrared camera measurement model. The result described above can have practical significance: that is, when measuring the temperature of high-emissivity objects, one should take into account that the measurement standard uncertainty (both absolute and relative) increases with the object temperature. An analogous situation occurs for the errors (Figure 4.4a).

Another interesting outcome of the combined uncertainty simulations is the observation that, even though the component of the relative combined standard uncertainty associated with object emissivity ε_{ob} is independent of ε_{ob}, the (whole) combined standard uncertainty (absolute as well as relative) depends on ε_{ob} – see Table 5.9. To make this dependence clearer, we increased the relative standard uncertainty associated with the object emissivity to 30%. Then, for $\varepsilon_{ob} = 0.9$ and $T_{ob} = 323$ K, the combined standard uncertainty is equal to 9.8 K, which is about 3% of the measured value. Next, the calculations were repeated for $\varepsilon_{ob} = 0.4$, keeping all other simulation parameters unchanged. The new combined standard uncertainty is 16 K, which is about 5% of the measured value. We must emphasize that in both the above cases the relative combined standard uncertainty associated with emissivity ε_{ob} was equal to 30%. We would not have come to this conclusion if we had looked only at components of the combined standard uncertainty (or components of the total error in Section 4.3). Thus, analysis of the uncertainty components alone is not enough to evaluate the influence of reference values of the input variables on the uncertainty of temperature measurement in infrared thermography. Additional simulations of the combined standard uncertainty are necessary.

Yet another important conclusion resulting from the above results refers to the evaluation of 95% coverage intervals. The method for the propagation of distributions and the Monte Carlo simulations allowed for the unique determination of the limits of these intervals. The assumption of the Gaussian probability distribution for the model output variable ensures safe results from the point of view of potential underestimation of the 95% coverage interval.

Summing up this chapter on the uncertainties of infrared thermography measurements, we can state that the concept of uncertainty makes it possible to complement our knowledge of the considered measurement model. In turn, propagation of the distributions allows us to estimate the combined standard uncertainty as well as coverage intervals of measurements in infrared thermography. In addition, analysis of uncertainties is a perfect tool for estimating measurement accuracy when statistical interactions occur.

Example 5.2 *Evaluation of the combined standard uncertainty $u_c(T_{ob})$ and the 95% coverage interval of infrared thermography measurement*

Let us consider a set of several thermograms recorded for an object whose surface has constant emissivity. The object temperature estimate is $T_{ob} = 363$ K. The emissivity measured using a contact method is $\varepsilon_{ob} = 0.9$, and the estimated relative standard uncertainty of the measurement is $u(\varepsilon_{ob}) = 10\%$. Relative standard uncertainties of the other input variables of model (3.17) are assumed as in Table 5.8, and estimates of these variables as in Table 5.7. Uniform probability distributions of all the input variables of model (3.17) are assumed for evaluation of the combined standard uncertainty.

The results presented in Table 5.9 are valid for a measurement defined in this way. For $T_{ob} = 363$ K and $\varepsilon_{ob} = 0.9$, the relative combined standard uncertainty $u_c(T_{ob}) = 1.5\%$. From Figure 5.60 and Table 5.9, we can see that the minimum width of the 95% coverage interval is equal to 18 K and that the limits of the coverage interval are $I_{95\%} = [355, 373]$.

6

Summary

The aim of this monograph was to present the issues associated with the estimation of errors and uncertainties of measurements in infrared thermography. Each non-contact temperature measurement (e.g. with an infrared camera) is a very specific measurement due, firstly, to a large number of influencing quantities and, secondly, to the high nonlinearities of a measurement model which result in inefficient analytic methods of evaluating accuracy. In this work, we proposed to apply the (exact) method of increments to evaluate the errors of temperature measurement by means of an infrared camera. At first, we presented the basic laws and definitions associated with infrared thermography measurements, namely radiative heat transfer (Section 2.2), the concept of emissivity (Section 2.3), as well as the principles of operation and basic metrological parameters of modern infrared cameras (Section 2.4). Next, we discussed the processing algorithm of a camera measurement path and the mathematical model of measurement in infrared thermography following from this algorithm (Section 3.2). The algorithm is described using the example of FLIR's ThermaCAM LW series 500 measurement camera, which is universal and typical of most thermal imaging systems now manufactured throughout the world.

Knowledge of the mathematical model of temperature measurement allowed us to determine the components of the error in the method. Analysis of the calculation results showed that the temperature measurement error depends mostly on components associated with the emissivity of the object ε_{ob} whose temperature is measured (Section 4.3.1). The second important quantity, in terms of influence on the measurement error, turned out to be the ambient temperature T_o. Our results also proved that the error components associated with the relative humidity ω, the camera-to-object distance d and the atmospheric temperature T_{atm} do not contribute significantly to the total error in the method (Sections 4.3.3–4.3.5 and 5.3.3–5.3.5). Analysis of the error of the method allowed us to investigate the model sensitivity to changes in the input variables. We would like to emphasize that classical error analysis takes into consideration only the influence of systematic interactions. Such interactions relate to strictly defined measurement conditions that are difficult to realize in practice. Therefore, in this monograph, we also investigated random interactions. This investigation is based on the concept of uncertainty of a processing algorithm (Chapter 5). There, we presented the methodology of simulation research of the combined standard uncertainty in infrared

Infrared Thermography: Errors and Uncertainties Waldemar Minkina and Sebastian Dudzik

thermography measurement. Components of the combined standard uncertainty were evaluated assuming that measurements of each model's input quantity can be represented by a random variable (Section 5.3). Parameters (i.e. standard deviations and expected values) of the input random variables as well as shapes of the probability density distributions used in simulations were defined in advance. We assumed one of two typical (in the theory of measurement) distributions: the uniform distribution, which describes the critical case; or the logarithmic Gaussian distribution, whose random variable has only positive values. The simulation results are presented as graphs of components of the combined standard uncertainty in infrared thermography measurement for several different estimates of the input variables (such as the object emissivity or the camera-to-object distance).

Investigations of components of the combined standard uncertainty were conducted first for uncorrelated input variables of the measurement model in infrared thermography. It turns out, however, that estimation of the influence of cross-correlations among the variables on measurement accuracy is an important issue. In our opinion, in general, some variables are correlated with each other (Section 5.4). This does not necessarily mean that interdependencies occur between physical input quantities. We should remember that the input random variables represent results of measurements of the model quantities. Analysis of the influence of correlations showed that this influence depends on measurement conditions and these may differ greatly. The simulation results proved that the combined standard uncertainty depends mostly on the correlation between variables representing the object emissivity ε_{ob} and the ambient temperature T_o (Section 5.4.2). The simulations showed that the correlation between variables representing the ambient temperature T_o and the atmospheric temperature T_{atm} affects the temperature measurement uncertainty to some extent as well. In general, we stated that neglecting the correlations between these variables may lead to overestimation or underestimation of the combined standard uncertainty of the object temperature $u_c(T_{ob})$.

As mentioned in Section 3.2, our main intention was to link the investigation of errors to the investigation of uncertainties, because they do not exclude each other. Moreover, they complement each other and extend our knowledge of the considered measurement model and, consequently, indicate how to avoid sources of inaccuracy. Therefore, Section 5.5 dealt with the investigation of the combined standard uncertainty in the infrared thermography measurement model. Our research had two main goals: firstly, evaluation of the combined standard uncertainty under different measurement conditions, in order to determine its dependence on individual input variables; and secondly, estimation of the 95% coverage intervals taking into account the actual probability distributions of the model output variable. Distributions of the output variable (measured object temperature) were evaluated using the propagation of distributions, proposed by Working Group No. 1 formed by the Common Committee for Basic Problems in Metrology of the International Bureau of Weights and Measures. The propagation of distributions, based on the Monte Carlo simulations, also allowed us to estimate the 95% coverage interval of the measurement in infrared thermography. One of the basic conclusions, drawn from analysis of the combined standard uncertainty, is that despite the actual asymmetry of the output variable distribution, it is safe to assume the Gaussian distribution for evaluation of the 95% coverage interval. It turned out that the propagation of distributions is a perfect tool for evaluating the combined standard uncertainty in infrared thermography as well as for coverage intervals associated with this uncertainty.

The subject matter of this monograph (i.e. estimation of accuracy of infrared thermography measurement) is extensive and encompasses numerous branches of technology. The monograph does not pretend to be a complete study of the subject. Such a study would have to include analyses of different models and measurement conditions, models of phenomena affecting the accuracy (such as atmospheric transmission), etc. On the other hand, such deep analyses would not be purposeful from a practical point of view. An extensive set of tables, characteristics and other detailed data would not necessarily make possible a better evaluation of measurement accuracy. Therefore, we concentrated on more general conclusions that enable practical applications of the presented results. We hope that the described methods of evaluating errors and uncertainties will help to improve infrared thermography measurement accuracy under real conditions. Yet the final test of any theory is experiment . . .

Appendix A

MATLAB Scripts and Functions

A.1 Typesetting of the Code

In this work, the accuracy of the infrared thermography measurements was investigated with sophisticated software, created in the MATLAB computational environment. In order to help users of the infrared camera to estimate the uncertainty of the non-contact temperature measurement for their own conditions, we present below the source code of the MATLAB programs applied in this book. It can be prepared by typesetting into the editor of the MATLAB environment or by digitizing with OCR (Optical Character Recognition) software. The m-files should be located in the same folder as the MATLAB environment (e.g. in Matlab\Work). The m-files were created as functions and scripts. Table A.1 lists the collected scripts, and Table A.2 the functions.

Presented below are the procedures for calculating the accuracy of infrared thermography measurements. In calculations it is necessary to supply the measurement parameters from the *.img file. For this reason, the first stage of the program is to input the name of this file (recorded with the infrared camera). This file must be located in the same directory as the suitable m-files. It must be emphasized that the plots generated with this software can differ from the results described in this book. This is a result of the different calibration and adjustment parameters and the conditions of infrared thermography measurement obtained by different infrared cameras.

A.2 Procedure for Calculating the Components of Combined Standard Uncertainty in Infrared Thermography Measurement Using the Presented Software

1. Type `components` in the MATLAB command window.
2. Enter the numerical data (measurement conditions, standard uncertainties, and so on) in accordance with the messages on the screen.
3. When the script is finished, type `plotcomponents` in order to plot the results of the calculations. As a result of the script executing, five graphs of suitable components will be plotted.

Table A.1 The m-file scripts used in calculations of the accuracy of infrared thermography measurements

Name of m-file	Description
`components.m`	Calculates components of the combined standard uncertainty of the object temperature for FLIR ThermaCAM infrared cameras
`correlations.m`	Simulates the influence of the cross-correlations between the input variables of the FLIR ThermaCAM infrared camera model on the combined standard uncertainty
`coverint.m`	Calculates the coverage interval for FLIR ThermaCAM infrared cameras
`plotcomponents.m`	Plots the component of the given input random variable with `components.m`
`plotcorrelations.m`	Plots the correlation between the cross-correlated input variables of the FLIR ThermaCAM infrared camera model
`plotcorrsens.m`	Plots the combined standard uncertainty vs. correlation coefficient between input variables of the FLIR ThermaCAM infrared camera model
`plotresults.m`	Plots results for calculations conducted with `coverint.m`

A.3 Procedure for Calculating the Coverage Interval and Combined Standard Uncertainty in Infrared Thermography Measurement Using the Presented Software

1. Type `coverint` in the MATLAB command window.
2. Enter the numerical data (measurement conditions, standard uncertainties, and so on) in accordance with the messages on the screen.

Table A.2 The m-file functions used in calculations of the accuracy of infrared thermography measurements

Name of m-file	Description
`cameramodel.m`	The model of the measurement with the ThermaCAM infrared camera
`distribute.m`	Calculates an approximation of distribution for the input random variable
`estlogpars.m`	Estimates the log-normal distribution parameters on the basis of expected value and variance
`estunifrpars.m`	Estimates the uniform distribution parameters on the basis of expected value and variance
`gencorrlog.m`	Generates two-dimensional, cross-correlated log-normal distribution
`gencorruni.m`	Generates two-dimensional, cross-correlated uniform distribution
`loadimgheader.m`	Loads a header of *.img file
`plotcorrelated.m`	Creates a plot of the cross-correlated input random variables for given value of the correlation coefficient
`plotdistcomp.m`	Plots the distribution of the input random variable
`plotsensitive.m`	Creates a plot for `plotcorrsens.m`
`readimgdatablock.m`	Reads single typed value from the *.img file
`temptosignal.m`	Calculates pixel value on the basis of temperature value

3. When the script is finished, type `plotresults` in order to plot the results of the calculations. As a result of the script executing, seven graphs of suitable components will be plotted. Five graphs present histograms of the input random variables, the sixth graph presents a probability density function of the output random variable and the seventh graph presents an approximation of the cumulative distribution function of the output variable.

A.4 Procedure for Simulating the Cross-correlations Between the Input Variables of the Infrared Camera Model Using the Presented Software

1. Type `correlations` in the MATLAB command window.
2. Enter the numerical data (measurement conditions, standard uncertainties, and so on) in accordance with the messages on the screen.
3. When the script is finished, type `plotcorrelations` in order to plot the cross-correlated input random variables. As a result of the script executing, a graph of a suitable pair of the cross-correlated input variables will be plotted.
4. Type `plotcorrsens` in order to plot the combined standard uncertainty vs. correlation coefficient between input variables of the infrared thermography measurement. As a result of the script executing, a graph will be plotted. It presents the combined standard uncertainty of the output variable vs. the correlation coefficient between input variables selected when the `correlations` script was executed.

A.5 MATLAB Source Code (Scripts)

```
%%%%%%%%%%%%%%%%%%%%%%%%%%%%%%%%%%%%%%%%%%%%%%%%%
%                                               %
% COMPONENTS.M                                  %
%                                               %
% Calculates components of the combined         %
% standard uncertainty of the object            %
% temperature for FLIR ThermaCAM infrared       %
% cameras with the method for the propagation   %
% of distribution with assumption of            %
% different types of distribution for input     %
% random variables                              %
%                                               %
%Copyright Feb, 2008 by Sebastian Dudzik        %
%(sebdud@el.pcz.czest.pl)                       %
%                                               %
%%%%%%%%%%%%%%%%%%%%%%%%%%%%%%%%%%%%%%%%%%%%%%%%%

% Initial screen
clc;
h={};
disp('****************************************');
disp('*                                      *');
disp('* Calculation of components of the     *');
disp('* combined standard uncertainty        *');
```

```
disp('*  of the object temperature              *');
disp('*  for FLIR ThermaCAM infrared cameras    *');
disp('*                                         *');
disp('******************************************');
disp(' ');

%- - - - - - - - - - - - - - - - - - - - - - - - - - - - - - - -
% Input data blocks
%- - - - - - - - - - - - - - - - - - - - - - - - - - - - - - - -

% Input for the name of file recorded with infrared camera
disp('*** FILE NAME AND MEASURED TEMPERATURE BLOCK ***');
disp(' ');
fileName=input(['Name of *.img recorded file:' ...
                '(must be in the same dir): '],'s');

% Reading the radiometric data from *.img file into the
% structure of h
% SEE ALSO: loadimgheader.m function
h = loadimgheader(fileName, h);

% Input for the temperature value of specified pixel
% in the thermogram
tObject=input('Value of measured temperature (K):');

disp(' ');
disp('*** REFERENCE CONDITIONS BLOCK ***');
disp(' ');

% Input for the emissivity value (set in the camera)
emiss=input('Value of emissivity: ');

% Input for the ambient temperature value
% (set in the camera)
tAmb=input('Value of ambient temperature (K): ');

% Input for the value of atmosphere temperature
% (set in the camera)
tAtm=input('Value of temperature of atmosphere (K): ');

% Input for the relative humidity value
% (set in the camera)
humRel=input('Value of relative humidity : ');

% Input for the camera-to-object distance value
% (set in the camera)
dist=input('Value of camera-to-object distance (m): ');

% Calculation of the signal-from-detector value
% on the basis tObject temperature
% SEE ALSO: temptosignal.m function
signal = temptosignal(tObject, emiss, tAtm, tAmb, humRel,dist,...
      h.alpha1, h.alpha2, h.beta1, h.beta2, h.X, h.R, ..
      h.B, h.F,h.obas, h.L, h.globalGain, h.globalOffset);
```

```
disp(' ');
disp(['*** RANGES OF THE STANDARD UNCERTAINTIES ' ...
    'OF INPUT VARIABLES BLOCK ***']);
disp(' ');

% Input for the range of the standard uncertainty of
% emissivity measurement
minEmissUn=input('Minimum uncertainty of emissivity (%): ');
maxEmissUn=input('Maximum uncertainty of emissivity (%): ');

% Input for the range of the standard uncertainty of
% ambient temperature measurement
minTAmbUn=input(['Minimum uncertainty of ' ...
                'ambient temperature (%): ']);
maxTAmbUn=input(['Maximum uncertainty of ' ...
                'ambient temperature (%): ']);

% Input for the range of the standard uncertainty of
% temperature of atmosphere measurement
minTAtmUn=input(['Minimum uncertainty of temperature' ...
                'of atmosphere (%): ']);
maxTAtmUn=input(['Maximum uncertainty of ' ...
                'temperature of atmosphere (%): ']);

% Input for the range of the standard uncertainty of relative
% humidity measurement
minHumRelUn=input(['Minimum uncertainty of ' ...
                  'relative humidity (%): ']);
maxHumRelUn=input(['Maximum uncertainty of ' ...
                  'relative humidity (%): ']);

% Input for the range of the standard uncertainty
% of camera-to object distance measurement
minDistUn=input(['Minimum uncertainty of ' ...
                'camera-to-object distance (%): ']);
maxDistUn=input(['Maximum uncertainty of ' ...
                'camera-to-object distance (%): ']);

% Input of the number of simulation points
nPoints = input('Number of simulation points: ');

% Calculation of the uncertainty values

% Emissivity
emissUn = linspace((minEmissUn*emiss)/100, ...
                  (maxEmissUn*emiss)/100,nPoints);
% Ambient temperature
tAmbUn = linspace((minTAmbUn*tAmb)/100, ...
                (maxTAmbUn*tAmb)/100,nPoints);

% Temperature of atmosphere
tAtmUn = linspace((minTAtmUn*tAtm)/100, ...
                (maxTAtmUn*tAtm)/100,nPoints);

% Relative humidity
```

```
humRelUn = linspace((minHumRelUn*humRel)/100, ...
                    (maxHumRelUn*humRel)/100,nPoints);

% Camera-to-object distance
distUn = linspace((minDistUn*dist)/100, ...
                  (maxDistUn*dist)/100,nPoints);

disp(' ');
disp(['*** THE DISTRIBUTIONS OF THE INPUT ' ...
     'RANDOM VARIABLES BLOCK ***']);
disp(' ');

disp('1 - Lognormal distribution');
disp('2 - Uniform distribution');
disp(' ');
typeOfDist = input('Enter type of distribution (1/2): ');

%-------------------------------
% Generation of the distributions
% of the input random variables
%-------------------------------

% Distribution for the input random variable
% representing emissivity
emissDistribution=zeros(1,10000);
if typeOfDist==1
    [a,b]=estlogpars(emiss, emissUn(1).^2);
    hlpVar=lognrnd(a,b,1,10000);
    emissDistribution=emissDistribution+hlpVar;
    for i=2:nPoints
        [a,b]=estlogpars(emiss, emissUn(i).^2);
        hlpVar=lognrnd(a,b,1,10000);
        emissDistribution=[emissDistribution; hlpVar];
    end;
    clear hlpVar ;
else
    [a,b]=estunifrpars(emiss, emissUn(1).^2);
    hlpVar=unifrnd(a,b,1,10000);
    emissDistribution=emissDistribution+hlpVar;
    for i=2:nPoints
        [a,b]=estunifrpars(emiss, emissUn(i).^2);
        hlpVar=unifrnd(a,b,1,10000);
        emissDistribution=[emissDistribution; hlpVar];
    end;
    clear hlpVar ;
end;

% Distribution for the input random variable representing
% ambient temperature
tAmbDistribution=zeros(1,10000);
if typeOfDist==1
    [a,b]=estlogpars(tAmb, tAmbUn(1).^2);
    hlpVar=lognrnd(a,b,1,10000);

    tAmbDistribution=tAmbDistribution+hlpVar;
```

```
    for i=2:nPoints
        [a,b]=estlogpars(tAmb, tAmbUn(i).^2);
        hlpVar=lognrnd(a,b,1,10000);
        tAmbDistribution=[tAmbDistribution; hlpVar];
    end;
    clear hlpVar ;
else
    [a,b]=estunifrpars(tAmb, tAmbUn(1).^2);
    hlpVar=unifrnd(a,b,1,10000);
    tAmbDistribution=tAmbDistribution+hlpVar;
    for i=2:nPoints
        [a,b]=estunifrpars(tAmb, tAmbUn(i).^2);
        hlpVar=unifrnd(a,b,1,10000);
        tAmbDistribution=[tAmbDistribution; hlpVar];
    end;
    clear hlpVar ;
end;

% Distribution for the input random variable representing
% temperature of atmosphere
tAtmDistribution=zeros(1,10000);
if typeOfDist==1
    [a,b]=estlogpars(tAtm, tAtmUn(1).^2);
    hlpVar=lognrnd(a,b,1,10000);
    tAtmDistribution=tAtmDistribution+hlpVar;
    for i=2:nPoints
        [a,b]=estlogpars(tAtm, tAtmUn(i).^2);
        hlpVar=lognrnd(a,b,1,10000);
        tAtmDistribution=[tAtmDistribution; hlpVar];
    end;
    clear hlpVar ;
else
    [a,b]=estunifrpars(tAtm, tAtmUn(1).^2);
    hlpVar=unifrnd(a,b,1,10000);
    tAtmDistribution=tAtmDistribution+hlpVar;
    for i=2:nPoints
        [a,b]=estunifrpars(tAtm, tAtmUn(i).^2);
        hlpVar=unifrnd(a,b,1,10000);
        tAtmDistribution=[tAtmDistribution; hlpVar];
    end;
    clear hlpVar ;
end;

% Distribution for the input random variable
% representing relative humidity
humRelDistribution=zeros(1,10000);
if typeOfDist==1
    [a,b]=estlogpars(humRel, humRelUn(1).^2);
    hlpVar=lognrnd(a,b,1,10000);
    humRelDistribution=humRelDistribution+hlpVar;
    for i=2:nPoints
        [a,b]=estlogpars(humRel, humRelUn(i).^2);
        hlpVar=lognrnd(a,b,1,10000);
        humRelDistribution=[humRelDistribution; hlpVar];
```

```
    end;
    clear hlpVar ;
else
    [a,b]=estunifrpars(humRel, humRelUn(1).^2);
    hlpVar=unifrnd(a,b,1,10000);
    humRelDistribution=humRelDistribution+hlpVar;
    for i=2:nPoints
        [a,b]=estunifrpars(humRel, humRelUn(i).^2);
        hlpVar=unifrnd(a,b,1,10000);
        humRelDistribution=[humRelDistribution; hlpVar];
    end;
    clear hlpVar ;
end;

% Distribution for the input random variable
% representing camera-to-object distance
distDistribution=zeros(1,10000);
if typeOfDist==1
    [a,b]=estlogpars(dist, distUn(1).^2);
    hlpVar=lognrnd(a,b,1,10000);
    distDistribution=distDistribution+hlpVar;
    for i=2:nPoints
        [a,b]=estlogpars(dist, distUn(i).^2);
        hlpVar=lognrnd(a,b,1,10000);
        distDistribution=[distDistribution; hlpVar];
    end;
    clear hlpVar ;
else
    [a,b]=estunifrpars(dist, distUn(1).^2);
    hlpVar=unifrnd(a,b,1,10000);
    distDistribution=distDistribution+hlpVar;
    for i=2:nPoints
        [a,b]=estunifrpars(dist, distUn(i).^2);
        hlpVar=unifrnd(a,b,1,10000);
        distDistribution=[distDistribution; hlpVar];
    end;
    clear hlpVar ;
end;

%- - - - - - - - - - - - - - - - - - - - - - - - - - - - - -
% Simulations for the components of the combined
% standard uncertainty of the object temperature
%- - - - - - - - - - - - - - - - - - - - - - - - - - - - - -

% Emissivity component
emissComponent = zeros(10000,nPoints);
  for i=1:nPoints
        emissComponent(:,i) = cameramodel(signal, ...
        emissDistribution(i,:), tAmb, tAtm, humRel, ...
        dist, h.alpha1, h.alpha2, h.beta1, h.beta2, ...
        h.X, h.R, h.B, h.F, h.obas, h.L, ...
        h.globalGain,h.globalOffset);
  end
emissStd = std(emissComponent);
emissStdRel = (emissStd/tObject)*100;
```

```
% Ambient temperature component
tAmbComponent = zeros(10000,nPoints);
 for i=1:nPoints
       tAmbComponent(:,i) = cameramodel(signal, emiss, ...
       tAmbDistribution(i,:), tAtm, humRel, ...
       dist, h.alpha1, h.alpha2, h.beta1, h.beta2, ...
       h.X, h.R, h.B, h.F, h.obas, h.L, h.globalGain,...
       h.globalOffset);
  end
tAmbStd = std(tAmbComponent);
tAmbStdRel = (tAmbStd/tObject)*100;

% Atmosphere temperature component
tAtmComponent = zeros(10000,nPoints);
  for i=1:nPoints
        tAtmComponent(:,i) = cameramodel(signal, emiss, ...
        tAmb, tAtmDistribution(i,:), humRel, dist, ...
        h.alpha1, h.alpha2, h.beta1, h.beta2, h.X, h.R, h.B, ...
        h.F, h.obas, h.L, h.globalGain, h.globalOffset);
  end
tAtmStd = std(tAtmComponent);
tAtmStdRel = (tAtmStd/tObject)*100;

% Relative humidity component
humRelComponent = zeros(10000,nPoints);
  for i=1:nPoints
        humRelComponent(:,i) = cameramodel(signal, emiss, ...
        tAmb, tAtm, humRelDistribution(i,:), dist, h.alpha1, ...
        h.alpha2, h.beta1, h.beta2, h.X, h.R, h.B, h.F, ...
        h.obas, h.L, h.globalGain, h.globalOffset);
  end
humRelStd = std(humRelComponent);
humRelStdRel = (humRelStd/tObject)*100;

% Camera-to-object distance component
distComponent = zeros(10000,nPoints);
  for i=1:nPoints
        distComponent(:,i) = cameramodel(signal, emiss, ...
        tAmb, tAtm, humRel, distDistribution(i,:), h.alpha1, ...
        h.alpha2, h.beta1, h.beta2, h.X, h.R, h.B, h.F, ...
     h.obas, h.L, h.globalGain, h.globalOffset);
  end
distStd = std(distComponent);
distStdRel = (distStd/tObject)*100;

% End of COMPONENTS
```

```
%                                                %
%   PLOTRESULTS.M                                %
%                                                %
%   Plots results for calculations conducted     %
%   with COVERINT.M                              %
%                                                %
```

```
%   WARNING: Run after COVERINT.M                          %
%                                                          %
%Copyright Feb, 2008 by Sebastian Dudzik                   %
%(sebdud@el.pcz.czest.pl)                                  %
%                                                          %
%%%%%%%%%%%%%%%%%%%%%%%%%%%%%%%%%%%%%%%%%%%%%%%%%%%%%%%%%%%%

% Plot histogram of the input random variable representing
% emissivity
figure;
hist(x1,45);
h1=get(gca,'Child');
set(h1,'FaceColor','none','EdgeColor','blue')
title({'Histogram of the input random variable'; ...
        'representing emissivity'});
xlabel('Emissivity \it{\epsilon_{ob}}')
ylabel('Size [samples]');
annotation('textbox',[0.5,0.5,0.3,0.1], 'BackgroundColor', ...
             'white', 'String',{['Expected value: ' ...
             num2str(mean(x1),3)]; ['Standard deviation: '...
             num2str(std(x1),3)]});

% Plot histogram of ambient temperature distribution
figure;
hist(x2,45);
h2=get(gca,'Child');
set(h2,'FaceColor','none','EdgeColor','blue')
title({'Histogram of the input random variable'; ...
        'representing ambient temperature'});
xlabel('Ambient temperature \it{T_{o}}\rm (K)')
ylabel('Size [samples]');
annotation('textbox',[0.5,0.5,0.3,0.1], 'BackgroundColor', ...
             'white', 'String',{['Expected value: ' ...
            num2str(mean(x2),3) ' K']; ['Standard deviation: '...
            num2str(std(x2),3) ' K']});

% Plot histogram of temperature of atmosphere distribution
figure;
hist(x3,45);
h3=get(gca,'Child');
set(h3,'FaceColor','none','EdgeColor','blue')
title({'Histogram of the input random variable'; ...
        'representing atmosphere temperature'})
xlabel('Atmosphere temperature \it{T_{0}}\rm (K)')
ylabel('Size [samples]');
annotation('textbox',[0.5,0.5,0.3,0.1], 'BackgroundColor', ...
             'white', 'String',{['Expected value: ' ...
           num2str(mean(x3),3) ' K']; ['Standard deviation: ' ...
           num2str(std(x3),3) ' K']});

% Plot histogram of relative humidity distribution
figure;
hist(x4,45);
h4=get(gca,'Child');
set(h4,'FaceColor','none','EdgeColor','blue')
```

```
title({'Histogram of the input random variable'; ...
       'representing relative humidity'})
xlabel('Relative humidity \it{\omega}')
ylabel('Size [samples]');
annotation('textbox',[0.5,0.5,0.3,0.1],
'BackgroundColor', ...
          'white', 'String',{['Expected value: ' ...
          num2str(mean(x4),3)]; ['Standard deviation: ' ...
          num2str(std(x4),3)]});

% Plot histogram of camera-to-object distance distribution
figure;
hist(x5,45);
h5=get(gca,'Child');
set(h5,'FaceColor','none','EdgeColor','blue')
title({'Histogram of the input random variable'; ...
       'representing camera-to-object distance'})
xlabel('camera-to-object distance \it{d}\rm m')
ylabel('Size [samples]');
annotation('textbox',[0.5,0.5,0.3,0.1], 'BackgroundColor', ...
          'white', 'String',{['Expected value: ' ...
    num2str(mean(x5),3)]; ['Standard deviation: ' ...
    num2str(std(x5),3)]});

% Plot histogram of temperature distribution
figure
[n,k]=hist(temperature,40);
nn=(40*n)./((max(temperature)-min(temperature))*1e6);
hbar = bar(k,nn,1);
grid on;
xlabel('\it{T_{ob}},\rm K');
ylabel('\it{g(T_{ob})}');
set(hbar,'FaceColor', 'white');
hold;
h1 = plot([tLow, tLow],[0 max(nn)]);
set(h1,'LineWidth',2);
set(h1,'Color','black');
h2 = plot([tHigh, tHigh],[0 max(nn)]);
set(h2,'LineWidth',2);
set(h2,'Color','black');
title({'Probability density function of the output random variable';...
       'representing object temperature'})
annotation('textbox',[0.5,0.5,0.3,0.1], 'BackgroundColor', ...
    'white', 'String',{['Expected value: ' ...
        num2str(mean(temperature),3)]; ['Standard deviation: ' ...
        num2str(std(temperature),3)]});

% Plot approximates of temperature distribution function
figure;
h7=plot(tDist(:,1),tDist(:,2),'b-');
set(gca,'XMinorTick','on');
set(gca,'YMinorTick','on');
title({'Approximation of the cumulative distribution function'; ...
       'for output random variable'})
xlabel('Temperature \it{T_{ob}}\rm K')
```

```
ylabel('\itG(T_{ob})');
hold;
line([tLow tLow],[0 1],'Color','red', 'LineWidth',2);
line([tHigh tHigh],[0 1],'Color','red', 'LineWidth',2);
legend('Approximation of cumulative distribution function',...
       '95% coverage interval','Location','best');

% End of PLOTRESULTS
```

```
%%%%%%%%%%%%%%%%%%%%%%%%%%%%%%%%%%%%%%%%%%%%%%%%
%                                              %
%   CORRELATIONS.M                             %
%                                              %
%   Simulates the influence of the             %
%   cross-correlations between the input       %
%   variables of the FLIR ThermaCAM infrared   %
%   camera model on the combined standard      %
%   uncertainty                                %
%                                              %
%Copyright Feb, 2008 by Sebastian Dudzik       %
%(sebdud@el.pcz.czest.pl)                      %
%                                              %
%%%%%%%%%%%%%%%%%%%%%%%%%%%%%%%%%%%%%%%%%%%%%%%%

% Initial screen
clc;
h={};
disp('*************************************');
disp('*                                   *');
disp('*   Simulates the influence of the  *');
disp('*   cross-correlations between the input *');
disp('*   variables of the FLIR ThermaCAM *');
disp('*   infrared camera model on the    *');
disp('*   combined standard uncertainty   *');
disp('*                                   *');
disp('*************************************');
disp(' ');

% Input for the name of file recorded with infrared camera
disp('*** FILE NAME AND MEASURED TEMPERATURE BLOCK ***');
disp(' ');
fileName=input(['Name of *.img recorded file:' ...
             ' (must be in the same dir): '],'s');

% Reading the radiometric data from *.img file
% into the structure of h SEE ALSO: loadimgheader.m function
h = loadimgheader(fileName, h);

% Input for the temperature value of specified pixel
% in the thermogram
tObject=input('Value of measured temperature (K): ');

disp(' ');
disp('*** REFERENCE CONDITIONS BLOCK ***');
```

```
disp(' ');

% Input for the emissivity value (set in the camera)
emiss=input('Value of emissivity: ');

% Input for the ambient temperature value (set in the camera)
tAmb=input('Value of ambient temperature (K): ');

% Input for the value of atmosphere temperature
% (set in the camera)
tAtm=input('Value of temperature of atmosphere (K): ');

% Input for the relative humidity value (set in the camera)
humRel=input('Value of relative humidity : ');

% Input for the camera-to-object distance value
% (set in the camera)
dist=input('Value of camera-to-object distance (m): ');

disp(' ');
disp('*** STANDARD UNCERTAINTIES OF INPUT VARIABLES BLOCK ***');
disp(' ');

% Input for the standard uncertainty of emissivity measurement
emissUn=input('Standard uncertainty of emissivity: ');

% Input for the standard uncertainty of ambient
% temperature measurement
tAmbUn=input(['Standard uncertainty of ambient ' ...
          'temperature (K): ']);

% Input for the standard uncertainty of temperature
% of atmosphere measurement
tAtmUn=input(['Standard uncertainty of temperature ' ...
          'of atmosphere (K): ']);

% Input for the standard uncertainty of relative
% humidity measurement
humRelUn=input('Standard uncertainty of relative humidity : ');

% Input for the standard uncertainty of camera-to object
% distance measurement
distUn=input(['Standard uncertainty of camera-to-object '...
          'distance (m): ']);

NOSAMPLES = 10000; % Number of the random samples of the
                   % input random variable

parsNorm{1} = [emiss emissUn];
parsNorm{2} = [tAmb tAmbUn];
parsNorm{3} = [tAtm tAtmUn];
parsNorm{4} = [humRel humRelUn];
parsNorm{5} = [dist distUn];

disp(' ');
```

```
disp('*** PARAMETERS OF CROSS-CORRELATIONS VECTOR BLOCK ***');
disp(' ');

% Input data for construction of cross-correlation vector
jBegin = input('Starting value of cross-correlation vector: ');
jStep = input('Step value into cross-correlation vector: ');
jEnd = input('Ending value of cross-correlation vector: ');
jCorrCoef = jBegin:jStep:jEnd;
jLen = length(jCorrCoef);

disp(' ');
disp(['*** THE DISTRIBUTIONS OF THE INPUT ' ...
      'RANDOM VARIABLES BLOCK ***']);
disp(' ');

disp('1 - Lognormal distribution');
disp('2 - Uniform distribution');
disp(' ');
typeOfDist = input('Enter type of distribution (1/2): ');

% Choice of distribution type
if typeOfDist == 1 % Log-normal distribution

  % Estimation of the Log-normal distribution parameters
  for i = 1:5
     [parsLog{i}(1,1) parsLog{i}(1,2)] = ...
        estlogpars(parsNorm{i}(1,1), parsNorm{i} (1,2)^2);
  end;

  % Generation of the input random variable of Log-normal
  % type
  for i = 1:5
     inputs{i} = lognrnd(parsLog{i}(1,1), ...
        parsLog{i}(1,2), NOSAMPLES, jLen);
  end;

  disp(' ');
  disp(['*** THE CROSS-CORRELATED INPUT RANDOM ' ...
     'VARIABLES BLOCK ***']);
  disp(' ');

  % Indexes of the input variables assigned for
  % cross-correlation
  disp('The List of the input variables'' index');
  disp(' ');
  disp(' Index | Input variable');
  disp('-------|-----------------');
  disp('  1    | Emissivity');
  disp('  2    | Ambient temperature');
  disp('  3    | Atmosphere temperature');
  disp('  4    | Relative humidity');
  disp('  5    | Camera-to-object distance');
  disp(' ');
  kPopup = input(['Enter the index of the first ' ...
                  'cross-correlated input variable: ']);
```

```
    lPopup = input(['Enter the index  of the second ' ...
                    'cross-correlated input variable: ']);

    % Auxiliary cell array of parameters for cross-correlated
    % variables
    pNorm{1} = [parsNorm{kPopup}(1,1) parsNorm{kPopup}(1,2)];
    pNorm{2} = [parsNorm{lPopup}(1,1) parsNorm{lPopup}(1,2)];
    pLog{1} = [parsLog{kPopup}(1,1) parsLog{kPopup}(1,2)];
    pLog{2} = [parsLog{lPopup}(1,1)    parsLog{lPopup}(1,2)];

    % Generation of the cross-correlated variables
    biCorrVariable = gencorrlog(pNorm, pLog, jCorrCoef, ...
                                NOSAMPLES);

else % Uniform distribution

    % Estimation of the Uniform distribution parameters
    for i = 1:5
       [parsUni{i}(1,1) parsUni{i}(1,2)] = ...
         estunifrpars(parsNorm{i}(1,1), parsNorm{i} (1,2)^2);
    end;

    % Generation of the input random variable of Uniform type
    for i = 1:5
       inputs{i} = unifrnd(parsUni{i}(1,1), ...
           parsUni{i}(1,2), NOSAMPLES, jLen);
    end;

    disp(' ');
    disp(['*** THE CROSS-CORRELATED INPUT RANDOM ' ...
       'VARIABLES BLOCK ***']);
    disp(' ');

    % Indexes of the input variables assigned for
    % cross-correlation
    disp('The List of the input variables'' index');
    disp(' ');
    disp(' Index | Input variable');
    disp('--------|-------------------');
    disp('   1    | Emissivity');
    disp('   2    | Ambient temperature');
    disp('   3    | Atmosphere temperature');
    disp('   4    | Relative humidity');
    disp('   5    | Camera-to-object distance');
    disp(' ');
    kPopup = input(['Enter the index of the first ' ...
                    'cross-correlated input variable: ']);
    lPopup = input(['Enter the index of the second ' ...
                    'cross-correlated input variable: ']);

    % Auxiliary cell array of parameters for cross-correlated
    % variables
    pNorm{1} = [parsNorm{kPopup}(1,1) parsNorm{kPopup}(1,2)];
    pNorm{2} = [parsNorm{lPopup}(1,1) parsNorm{lPopup}(1,2)];
    pUni{1} = [parsUni{kPopup}(1,1) parsUni{kPopup}(1,2)];
```

```
   pUni{2} = [parsUni{lPopup}(1,1) parsUni{lPopup}(1,2)];

   % Generation of the cross-correlated variables
   biCorrVariable = gencorruni(pNorm, pUni, jCorrCoef, ...
                               NOSAMPLES);

end % End of the choice of the distribution type
   % Conversion of the cell array variables into the matrices
   for i=1:jLen
     kPopupVariable(:,i) = biCorrVariable{1,i}(:,1);
     lPopupVariable(:,i) = biCorrVariable{1,i}(:,2);
   end;

% Place the cross-correlated variables into the cell array
% of the input random variables
inputs{kPopup} = kPopupVariable;
inputs{lPopup} = lPopupVariable;

% Simulation of the output random variable (temperature)
signal = temptosignal(tObject, emiss, tAmb, tAtm,...
                humRel, dist, h.alpha1, h.alpha2,...
                h.beta1, h.beta2, h.X, h.R, h.B, ...
                h.F, h.obas, h.L, h.globalGain, ...
                h.globalOffset);

for i=1:jLen
    tOut(:,i) = cameramodel(signal, inputs{1} (:,i),...
                inputs{2}(:,i), inputs{3}(:,i),...
                inputs{4}(:,i), inputs{5}(:,i),...
                h.alpha1, h.alpha2, h.beta1, ...
                h.beta2, h.X, h.R, h.B, h.F, ...
                h.obas, h.L, h.globalGain, ...
                h.globalOffset);
end;

% End of CORRELATIONS.M
```

```
%%%%%%%%%%%%%%%%%%%%%%%%%%%%%%%%%%%%%%%%%%%%%%%%%%%%
%                                                  %
%    COVERINT.M                                    %
%                                                  %
%    Calculates coverage interval for FLIR         %
%    ThermaCAM infrared cameras with the method    %
%    for the propagation of distribution           %
%    assuming uniform distribution of input        %
%    values                                        %
%                                                  %
%Copyright Feb, 2008 by Sebastian Dudzik           %
%(sebdud@el.pcz.czest.pl)                          %
%                                                  %
%%%%%%%%%%%%%%%%%%%%%%%%%%%%%%%%%%%%%%%%%%%%%%%%%%%%

% Initial screen
clc;
```

```
h={};
disp('****************************************');
disp('*                                      *');
disp('*  Calculation of coverage interval    *');
disp('*  for FLIR ThermaCAM infrared cameras *');
disp('*                                      *');
disp('****************************************');
disp(' ');

% Input for the name of file recorded with infrared camera
disp('*** FILE NAME AND MEASURED TEMPERATURE BLOCK ***');
disp(' ');
fileName=input(['Name of *.img recorded file:' ...
                '(must be in the same dir):'],'s');

% Reading the radiometric data from *.img file
% into the structure of h SEE ALSO: loadimgheader.m function
h = loadimgheader(fileName, h);

% Input for the temperature value of specified pixel
% in the thermogram
tObject=input('Value of measured temperature (K):');

disp(' ');
disp('*** REFERENCE CONDITIONS BLOCK ***');
disp(' ');

% Input for the emissivity value (set in the camera)
emiss=input('Value of emissivity: ');

% Input for the ambient temperature value (set in the camera)
tAmb=input('Value of ambient temperature (K): ');

% Input for the value of atmosphere temperature
% (set in the camera)
tAtm=input('Value of temperature of atmosphere (K): ');

% Input for the relative humidity value (set in the camera)
humRel=input('Value of relative humidity : ');

% Input for the camera-to-object distance value
% (set in the camera)
dist=input('Value of camera-to-object distance (m): ');

disp(' ');
disp('*** STANDARD UNCERTAINTIES OF INPUT VARIABLES BLOCK ***');
disp(' ');

% Input for the standard uncertainty of emissivity measurement
emissUn=input('Standard uncertainty of emissivity: ');

% Input for the standard uncertainty of ambient
% temperature measurement
tAmbUn=input(['Standard uncertainty of ambient ' ...
              'temperature (K): ']);
```

```
% Input for the standard uncertainty of temperature
% of atmosphere measurement
tAtmUn=input(['Standard uncertainty of temperature ' ...
              'of atmosphere (K): ']);

% Input for the standard uncertainty of relative
% humidity measurement
humRelUn=input('Standard uncertainty of relative humidity : ');

% Input for the standard uncertainty of camera-to object
% distance measurement
distUn=input(['Standard uncertainty of camera-to-object '...
              'distance (m): ']);
% Calculation of the signal-from-detector value on the basis
% tObject temperature
% SEE ALSO: temptosignal.m function
signal = temptosignal(tObject, emiss, tAmb, tAtm, humRel,dist,...
         h.alpha1, h.alpha2, h.beta1, h.beta2, h.X, h.R, ...
         h.B, h.F,h.obas, h.L, h.globalGain, h.globalOffset);

% Estimation of parameters of the uniform distribution
% for the emissivity input variable
% SEE ALSO: estunifrpars.m function
[a1 b1]= estunifrpars(emiss, emissUn^2);

% Estimation of parameters of the uniform distribution
% for the temperature of atmosphere input variable
% SEE ALSO: estunifrpars.m function
[a2 b2]= estunifrpars(tAtm,tAtmUn^2);

% Estimation of parameters of the uniform distribution
% for the ambient temperature input variable
% SEE ALSO: estunifrpars.m function
[a3 b3]= estunifrpars(tAmb, tAmbUn^2);

% Estimation of parameters of the uniform distribution
% for the relative humidity input variable
% SEE ALSO: estunifrpars.m function
[a4 b4]= estunifrpars(humRel, humRelUn^2);

% Estimation of parameters of the uniform distribution
% for the camera-to-object distance input variable
% SEE ALSO: estunifrpars.m function
[a5 b5]= estunifrpars(dist, distUn^2);

% Generation of random uniform distribution of the
% emissivity input variable according to parameters
% calculated above (1e6 samples)
x1=unifrnd(a1,b1,1e6,1);

% Generation of random uniform distribution of the
% temperature of atmosphere input variable according to
% parameters calculated above (1e6 samples)
x2=unifrnd(a2,b2,1e6,1);
```

```
% Generation of random uniform distribution of the
% ambient temperature input variable according to
% parameters calculated above (1e6 samples)
x3=unifrnd(a3,b3,1e6,1);

% Generation of random uniform distribution of the relative
% humidity input variable according to parameters calculated
% above (1e6 samples)
x4=unifrnd(a4,b4,1e6,1);

% Generation of random uniform distribution of the
% camera-to-object distance input variable according
% to parameters calculated above (1e6 samples)
x5=unifrnd(a5,b5,1e6,1);

% Applying the method for the propagation of
% distribution for obtaining the temperature distribution
% SEE ALSO: cameramodel.m function
temperature = cameramodel(signal, x1, x2, x3, x4, x5,...
    h.alpha1, h.alpha2, h.beta1, h.beta2, h.X, h.R, ...
    h.B, h.F, h.obas, h.L, h.globalGain, h.globalOffset);

disp(' ');
disp('*** RESULTS BLOCK ***')
disp(' ');
disp('Combined standard uncertainty of object temperature');

% Calculating combined standard uncertainty of temperature
% of specified pixel
std(temperature)

% Calculation of 95% coverage interval for measured temperature
% SEE ALSO: distribute.m function
tDist = distribute(temperature);
i=find(tDist(:,2)<=0.025);
tLow = tDist(i(end),1);
i=find(tDist(:,2)<=0.975);
tHigh = tDist(i(end),1);
disp(' ');
disp(['95% coverage interval ([tLow tHigh]): [' ...
    num2str(tLow) ', ' num2str(tHigh) ']']);

% End of COVERINT.M
```

```
%%%%%%%%%%%%%%%%%%%%%%%%%%%%%%%%%%%%%%%%%%%%%%%
%                                             %
%   PLOTCOMPONENTS.M                          %
%                                             %
%   Plots component of the given input random %
%   variable with COMPONENTS.M                %
%                                             %
%   WARNING: Run after COMPONENTS.M           %
%                                             %
%Copyright Feb, 2008 by Sebastian Dudzik      %
```

```
%(sebdud@el.pcz.czest.pl)                           %
%                                                   %
%%%%%%%%%%%%%%%%%%%%%%%%%%%%%%%%%%%%%%%%%%%%%%%%%%%%%

% Plot the emissivity component
plot(100*(emissUn/emiss),emissStdRel);
xlabel(['Standard uncertainty of the emissivity ' ...
       'u[\epsilon_{ob}]\rm (%)']);
ylabel(['Component of the standard uncertainty ' ...
       'u[T_{ob}]\rm (%)']);
set(gca,'XMinorTic','on');
set(gca,'YMinorTic','on');
grid on;
legend(['T_{ob} = ' num2str(tObject) ' K']);

% Plot the ambient temperature component
figure
plot(100*(tAmbUn/tAmb),tAmbStdRel);
xlabel(['Standard uncertainty of the ambient temperature' ...
       ' u[T_{o}]\rm (%)']);
ylabel(['Component of the standard uncertainty ' ...
       'u[T_{ob}]\rm (%)']);
set(gca,'XMinorTick','on');
set(gca,'YMinorTick','on');
grid on;
legend(['T_{ob} = ' num2str(tObject) ' K']);

% Plot the atmosphere temperature component
figure
plot(100*(tAtmUn/tAtm),tAtmStdRel);
xlabel(['Standard uncertainty of the atmosphere ' ...
       'temperature u[T_{atm}]\rm (%)']);
ylabel(['Component of the standard uncertainty' ...
       ' u[T_{ob}]\rm (%)']);
set(gca,'XMinorTick','on');
set(gca,'YMinorTick','on');
grid on;
legend(['T_{ob} = ' num2str(tObject) ' K']);
% Plot the relative humidity component

figure
plot(100*(humRelUn/humRel),humRelStdRel);
xlabel(['Standard uncertainty of the relative
humidity' ...
       ' u[\omega]\rm (%)']);
ylabel(['Component of the standard uncertainty' ...
       ' u[T_{ob}]\rm (%)']);
set(gca,'XMinorTick','on');
set(gca,'YMinorTick','on');
grid on;
legend(['T_{ob} = ' num2str(tObject) ' K']);

% Plot the camera-to-object distance component
figure
plot(100*(distUn/dist),distStdRel);
xlabel(['Standard uncertainty of the ' ...
```

```
        'camera-to-object distance u[d]\rm (%)']);
ylabel(['Component of the standard uncertainty' ...
        ' u[T_{ob}]\rm (%)']);
set(gca,'XMinorTick','on');
set(gca,'YMinorTick','on');
grid on;
legend(['T_{ob} = ' num2str(tObject) ' K']);

% End of PLOTCOMPONENTS.M
```

```
%%%%%%%%%%%%%%%%%%%%%%%%%%%%%%%%%%%%%%%%%%%%%%
%                                            %
%    PLOTCORRELATIONS.M                      %
%                                            %
%    Plots correlation between the           %
%    cross-correlated input variables of the %
%    FLIR ThermaCAM infrared camera model    %
%                                            %
%    WARNING: Run after CORRELATIONS.M       %
%                                            %
%See also: PLOTCORRELATED                    %
%                                            %
%Copyright Feb, 2008 by Sebastian Dudzik     %
%(sebdud@el.pcz.czest.pl)                    %
%                                            %
%%%%%%%%%%%%%%%%%%%%%%%%%%%%%%%%%%%%%%%%%%%%%%

% Initial screen
clc;
h={};
disp('**************************************');
disp('*                                    *');
disp('*   Plots correlations between the   *');
disp('*   cross-correlated input variables of  *');
disp('*   the FLIR ThermaCAM infrared camera  *');
disp('*   model                            *');
disp('*                                    *');
disp('**************************************');
disp(' ');

% *** CORRELATION COEFFICIENT INPUT BLOCK ***
disp(' ');
disp('The value of the correlation coefficient')
disp('for cross-correlated variables ');
disp(' ');
corrCoef = input('Enter value: ');

% *** DESIRED VARIABLE CORRELATION PLOT ***
firstVar = inputs{kPopup};
secondVar = inputs{lPopup};
diffCorrCoef=abs(jCorrCoef-corrCoef);
[y,i]=min(diffCorrCoef);
corrCoef = jCorrCoef(i);
[n1 ctr1] = hist(firstVar(:, i), 30);
[n2 ctr2] = hist(secondVar(:, i), 30);
```

```
plotcorrelated(firstVar(:, i), ...
              secondVar(:, i),corrCoef);

% Choice of appropriate input variable name
% for the x axis label
switch kPopup
    case 1
        xLab = 'Emissivity \it{\epsilon_{ob}}';
    case 2
        xLab = ['Ambient temperature' ...
              ' \it T_{o} \rm(K)'];
    case 3
        xLab = ['Temperature of atmosphere' ...
              ' \it T_{atm} \rm(K)'];
    case 4
        xLab = ['Relative humidity' ...
              ' \it{\omega}'];
    case 5
        xLab = ['Camera-to-object distance '...
              '\it{\d}'];
end

% Choice of appropriate input variable name
% for the y axis label
switch lPopup
    case 1
        yLab = 'Emissivity \it{\epsilon_{ob}}';
    case 2
        yLab = ['Ambient temperature' ...
              ' \it T_{o} \rm(K)'];
    case 3
        yLab = ['Temperature of atmosphere' ...
              ' \it T_{atm} \rm(K)'];
    case 4
        yLab = ['Relative humidity' ...
              ' \it{omega}'];
    case 5
        yLab = ['Camera-to-object distance '...
              '\it{d}'];
end

xlabel(xLab);
ylabel(yLab);

% End of PLOTCORRELATIONS.M
```

```
%%%%%%%%%%%%%%%%%%%%%%%%%%%%%%%%%%%%%%%%%%%
%                                         %
%   PLOTCORRSENS.M                        %
%                                         %
%   Plots the combined standard uncertainty vs.   %
%   correlation coefficient between input         %
%   variables of the FLIR ThermaCAM infrared      %
%   camera model                          %
%                                         %
```

```
%   WARNING: Run after CORRELATIONS.M                  %
%                                                      %
%See also: PLOTSENSITIVE                               %
%                                                      %
%Copyright Feb, 2008 by Sebastian Dudzik               %
%(sebdud@el.pcz.czest.pl)                              %
%                                                      %
%%%%%%%%%%%%%%%%%%%%%%%%%%%%%%%%%%%%%%%%%%%%%%%%%%%%%%%%

sensitive = 100.*std(tOut)./tObject;

plotsensitive(jCorrCoef, sensitive);

% End of PLOTCORRSENS.M
```

A.6 MATLAB Source Code (Functions)

```
function y = cameramodel(pixel, emissivity, tAmb, tAtm, ...
             humidity, distance, alpha1, alpha2, ...
             beta1, beta2, X, R, B, F, ...
             obas, L, globalGain, globalOffset)
%CAMERAMODEL The model of the measurement with ThermaCAM PM595
%            infrared camera
%
%   Y = CAMERAMODEL(PIXEL, EMISSIVITY, TAMB, TATM, HUMIDITY,
%    DISTANCE, ALPHA1, ALPHA2, BETA1, BETA2, X, R, B, F,
%   OBAS, L, GLOBALGAIN, GLOBALOFFSET)
%
%   TEMPERATURE: Measured temperature
%   EMISSIVITY:  Emissivity of object
%   TAMB:        Ambient temperature
%   TATM:        Temperature of atmosphere
%   HUMIDITY:    Relative humidity
%   DISTANCE:    Camera-to-object distance,
%
%   R, B, F, X,
%   ALPHA1, ALPHA2,
%   BETA1, BETA2,
%   GLOBALGAIN, GLOBALOFFSET,
%   L, OBAS: Calibration and adjusting parameters
%
%See also: LOADIMGHEADER, TEMPTOSIGNAL
%
%Copyright Feb, 2008 by Sebastian Dudzik
%(sebdud@el.pcz.czest.pl)

tAtmCelsius = tAtm-273.15;
h2o = humidity.*exp(1.5587+(6.939e-2).*tAtmCelsius - ...
(2.7816e-4).*tAtmCelsius.^ 2 + (6.8455e-7).* tAtmCelsius.^3);

% Calculation of transmission of atmosphere
```

```
tau = X.*exp((-1).*sqrt(distance).*(alpha1 + ...
beta1.*sqrt(h2o)))+(1 - X).*exp((-1).*...
sqrt(distance).*(alpha2 + beta2.*sqrt(h2o)));

% Linearization of pixel values
lFunc = (pixel - obas)./(1 - (L.*(pixel - obas)));

% Calculation of compensated pixel values
absPixel = globalGain.*lFunc + globalOffset;

% Calculation of detector signal value
k1 = 1./(emissivity.*tau);
k2 = (((1 - emissivity)./emissivity).*...
(R./(exp(B./tAmb) - F)))+(((1 -tau)./ ...
(emissivity.*tau)).*(R./(exp(B./tAmb) - F)));

objSignal = (k1.*absPixel/2)-k2;

% Calculation of temperature values
y = B./log((R./objSignal)+F);

% End of CAMERAMODEL
```

```
function [y] = distribute(inputVector);
%DISTRIBUTE Calculates the approximation of distribution for the input
%          random variable INPUTVECTOR
%
%   Y = DISTRIBUTE(INPUTVECTOR)
%   INPUTVECTOR: Input random variable
%
%See also: TEMPTOSIGNAL, CAMERAMODEL
%
%Copyright Feb, 2008 by Sebastian Dudzik.
%(sebdud@el.pcz.czest.pl)
inputSorted = sort(inputVector);
M = length(inputVector);
pr = zeros(1,M);
i = 1:M;
pr=(i-0.5)./M;
y = [inputSorted pr'];

% End of DISTRIBUTE
```

```
function [miLog, sigmaLog] = estlogpars(expectedVal, variance)
%ESTLOGPARS estimates of parameters of the log-normal
%         distribution on the basis of expected value and
%        variance
%
%   [MILOG, SIGMALOG] = ESTLOGPARS(EXPECTEDVAL, VARIANCE)
%
%   EXPECTEDVAL : Expected value
%   VARIANCE : Variance
%   MILOG, SIGMAL0G : parameters of the log-normal
%                     distribution
%
```

```
%See also: ESTUNIFRPARS

sigmaLog = sqrt(log((variance + expectedVal.^2)./expectedVal.^2));
miLog = log(expectedVal.^2./sqrt(variance +expectedVal.^2));

% End of ESLOGPARS
```

```
function [a, b] = estunifrpars(expectedVal, variance)
%ESTUNIFRPARS Estimates of parameters of uniform
%            distribution on the basis of the
%            expected value and variance
%
%   EXPECTEDVAL : Expected value
%   VARIANCE : Variance
%   A, B : Parameters of the uniform distribution
%
%See also: ESTLOGPARS

a = expectedVal - sqrt(3 * variance);
b = expectedVal + sqrt(3 * variance);

% End of ESTRUNIFRPARS.M
```

```
function biLogVariable = gencorrlog(parsNorm, parsLog, ...
                            jCorrCoef, noSamples);
%GENCORRLOG Generates 2-dimensional, cross-correlated
%          log-normal distribution
%   BIUNIVARIABLE = GENCORRLOG(parsNorm, parsLog,
%                               jCorrCoef, noSamples);
%
%   parsNorm : Cell array with parameters of the input
%              variables of the normal distribution
%              {[mu1 sig1] [mu2 sig2]}
%   parsLog : Cell array with parameters of the input
%                variables of the log-normal distribution
%              {[mu1 sig1] [mu2 sig2]}
% jCorrCoef : Vector of the cross-correlation coefficients
%               for input variables
%
%   noSamples : Number of samples in the resulted log-normal
%               distributions
%
%See also: GENCORRUNI
%Copyright Feb, 2008 by Sebastian Dudzik.
%(sebdud@el.pcz.czest.pl)

jLen = length(jCorrCoef);
% Covariance matrices of the input normal distributions
for i = 1:jLen;
  sigmaDep{i} = [parsNorm{1}(1,2)^2
jCorrCoef(i)*parsNorm{1}(1,2)*...
             parsNorm{2}(1,2);
jCorrCoef(i)*parsNorm{1}(1,2)*...
              parsNorm{2}(1,2) parsNorm{2}(1,2)^2];
```

```
end;
% 2-dimensional variable of normal distribution
for i = 1:jLen
   biNormalVariable{i} = mvnrnd([parsNorm{1}(1,1) parsNorm{2}(1,1)], ...
                                  sigmaDep{i}, noSamples);
end;
% Normalized variable (copula) of normal distribution
for i = 1:jLen
   biHelpVariable{i}=[normcdf(biNormalVariable{i}(:,1),
parsNorm{1}(1,1),...
                  parsNorm{1}(1,2)) ...
                     normcdf(biNormalVariable{i}(:,2),
parsNorm{2}(1,1),...
                  parsNorm{2}(1,2))];
end;
% 2-dimensional, cross-correlated variable of log-normal distribution
for i=1:jLen
   biLogVariable{i} = [logninv(biHelpVariable{i}(:,1), parsLog{1}(1,1), ...
                  parsLog{1}(1,2)) ...
                  logninv(biHelpVariable{i}(:,2),
parsLog{2}(1,1), ...
                   parsLog{2}(1,2))];
end;

% End of GENCORRLOG.M
```

```
function biUniVariable = gencorruni(parsNorm, ...
                                   parsUni, jCorrCoef, noSamples);
%GENCORRUNI Generates 2-dimensional, cross-correlated
%              uniform distribution
%   BIUNIVARIABLE = GENCORRUNI(parsNorm, parsUni,
%                                 jCorrCoef, noSamples);
%
%   parsNorm : Cell array with parameters of the input
%              variables of the normal distribution
%              {[mu1 sig1] [mu2 sig2]}
%   parsUni : Cell array with parameters of the input
%              variables of the uniform distribution
%              {[mu1 sig1] [mu2 sig2]}
%   jCorrCoef : Vector of the cross-correlation coefficients
%               for input variables
%
%   noSamples : Number of samples in the resulted uniform
%              distributions
%
%See also: GENCORRLOG
%Copyright Feb, 2008 by Sebastian Dudzik.
%(sebdud@el.pcz.czest.pl)

jLen = length(jCorrCoef);
% Covariance matrices of the input normal distributions
for i = 1:jLen;
   sigmaDep{i} = [parsNorm{1}(1,2)^2 jCorrCoef(i)*...
                 parsNorm{1}(1,2)* parsNorm{2}(1,2);...
```

```
                jCorrCoef(i)*parsNorm{1}(1,2)*...
                parsNorm{2}(1,2) parsNorm{2}(1,2)^2];
end;
% 2-dimensional variable of normal distribution
for i = 1:jLen
  biNormalVariable{i} = mvnrnd([parsNorm{1}(1,1) ...
                               parsNorm{2}(1,1)], ...
                               sigmaDep{i}, noSamples);
end;
% Normalized variable (copula) of normal distribution
for i = 1:jLen
   biHelpVariable{i}=[normcdf(biNormalVariable{i}(:,1), ...
                        parsNorm{1}(1,1),...
                        parsNorm{1}(1,2)) ...
                        normcdf(biNormal Variable{i}(:,2), ...
                        parsNorm{2}(1,1),...
                        parsNorm{2}(1,2))];
end;
% 2-dimensional, cross-correlated variable of uniform
% distribution
for i=1:jLen
  biUniVariable{i} = [unifinv(biHelpVariable{i}(:,1), ...
                        parsUni{1}(1,1), ...
                        parsUni{1}(1,2)) ...
                        unifinv(biHelp Variable{i}(:,2), ...
                        parsUni{2}(1,1), ...
                        parsUni{2}(1,2))];
end;

% End of GENCORRUNI.M
```

```
function h = loadimgheader(file, h)
%LOADIMGHEADER Loads a header of *.img file
%  H = LOADIMGHEADER(FILE, H) function returns the
%                             measurement parameters
%                             into the H structure
%  H = {
%         R, B, F, X, alpha1, alpha2,
%         beta1, beta2, tMin, tMax: float32;
%         globalGain, L: float32;
%         globalOffset: int32;
%         obas: uint16;
%         emissivity, distance, tAmbient, tAtmosphere,
%         humidity: float32;
%      }
%See also: TEMPTOSIGNAL, CAMERAMODEL
%
%Copyright Feb, 2008 by Sebastian Dudzik
%(sebdud@el.pcz.czest.pl)

% Addresses of the start of the header reading
index=readimgdatablock(file,0,'18',0,'lu');
block=readimgdatablock(file,0,index+hex2dec('0c'),1,'lu');
% Calibration parameters
```

```
calBlock=block+hex2dec('44');
h.R=readimgdatablock(file,calBlock,'0',0,'f');
h.B=readimgdatablock(file,calBlock,'4',0,'f');
h.F=readimgdatablock(file,calBlock,'8',0,'f');
h.X=readimgdatablock(file,calBlock,'1c',0,'f');
h.alpha1=readimgdatablock(file,calBlock,'c',0,'f');
h.alpha2=readimgdatablock(file,calBlock,'10',0,'f');
h.beta1=readimgdatablock(file,calBlock,'14',0,'f');
h.beta2=readimgdatablock(file,calBlock,'18',0,'f');
h.tMax=readimgdatablock(file,calBlock,'20',0,'f');
h.tMin=readimgdatablock(file,calBlock,'24',0,'f');

% Adjusting parameters
adjustBlock=block+hex2dec('10c');
h.globalGain=readimgdatablock(file,adjustBlock,'4',0,'f');
h.obas=readimgdatablock(file,adjustBlock,'1a',0,'s');
h.globalOffset=readimgdatablock(file,adjustBlock,'0',0,'l');
h.L=readimgdatablock(file,adjustBlock,'10',0,'f');

% Object parameters
objBlock=block+hex2dec('2c');
h.emissivity=readimgdatablock(file,objBlock,'0',0,'f');
h.distance=readimgdatablock(file,objBlock,'4',0,'f');
h.tAmbient=readimgdatablock(file,objBlock,'8',0,'f');
h.tAtmosphere=readimgdatablock(file,objBlock,'10',0,'f');
h.humidity=readimgdatablock(file,objBlock,'14',0,'f');

% End of LOADIMGHEADER
```

```
function plotcorrelated(x, y,i)
%PLOTCORRELATED(X,Y,I)
%   X: First cross-correlated input variable
%   Y: Second cross-correlated input variable
%   I: Value of the cross-correlation coefficient
%
%Copyright Feb, 2008 by Sebastian Dudzik.
%(sebdud@el.pcz.czest.pl)

% Create plot
figure;
plot1 = plot(...
   x,y,...
   'LineStyle','none',...
   'Marker','.',...
   'MarkerSize',4);

text(min(x)+0.01*min(x), max(y) - ...
     0.05*(max(y) - min(y)), ...
     ['\rho = ' num2str(i)], 'FontName', ...
     'Arial Unicode MS', 'FontSize', ...
     12, 'FontWeight', 'bold');

% End of PLOTCORRELATED.M
```

```
function [y]=plotdistcomp(distribution, nOfSimPoint);
%PLOTDISTCOMP Plots distribution of the input random variable
%
%   [Y] = PLOTDISTCOMP(DISTRIBUTION, NOFSIMPOINT)
%
%   DISTRIBUTION: distribution of the input random variable
%   NOFSIMPOINT: Number of simulation point
%
%   WARNING: Run after COMPONENTS.M
%
%Copyright Feb, 2008 by Sebastian Dudzik
%(sebdud@el.pcz.czest.pl)

% Plot histogram of the component
hist(distribution(nOfSimPoint,:),20); figure(gcf);

% End of PLOTDISTCOMP.M
```

```
function plotsensitive(x, y)
%PLOTSENSITIVE(X,Y)
%   X: vector of x data
%   Y: matrix of y data
%
%Copyright Feb, 2008 by Sebastian Dudzik.
%(sebdud@el.pcz.czest.pl)

% Create figure
figure1 = figure('PaperPosition', ...
                 [0.634 6.345 18.79 15.23], ...
                 'PaperSize',[20.98 29.68],...
                 'Position',[176 222 658 511]);

% Create axes
axes1 = axes(...
   'FontSize',10,...
   'XGrid','on',...
   'YGrid','on',...
   'Parent',figure1);
xlim(axes1,[-0.99 0.99]);
hx = xlabel(axes1,'\it{\rho}');
hy = ylabel(axes1,'u_c(T_{ob}), %');
box(axes1,'on');
hold(axes1,'all');

% Create multiple lines using matrix input to plot
plot1 = plot(x, y);

% End of PLOTSENSITIVE.M
```

```
function y = readimgdatablock(name, start, adress, ...
                              typeOfAdress, typeOfData)
%READIMGDATABLOCK Reads single typed value from the *.img file
```

```
%   Y = READIMGDATABLOCK(NAME, START, ADRESS,
%                        TYPEOFADRESS, TYPEOFDATA)
%
%   NAME         - File name
%   START        - Starting block address
%   ADRESS       - Data address relative to
                   starting block
%                  address
%   TYPEOFADRESS - Data address type:
                   (0) - hexadecimal,
%                  (1) - decimal
%     TYPEOFDATA - Data type: 'f' - float32, 'lu' - uint32,
%                         'l' - int32, 's' - uint16
%
%Copyright Feb, 2008 by Sebastian Dudzik.

fid = fopen(name,'r','b');
if typeOfAdress == 0
   adres = start+hex2dec(adress);
else
   adres=adress+start;
end

if typeOfData == 'f'

   status = fseek(fid,adres,'bof');
   [y,count] = fread(fid,1,'float32');

elseif typeOfData == 'lu'

   status = fseek(fid,adres,'bof');
   [y,count] = fread(fid,1,'uint32');

elseif typeOfData == 'l'

   status = fseek(fid,adres,'bof');
   [y,count] = fread(fid,1,'int32');

elseif typeOfData == 's'

   status = fseek(fid,adres,'bof');
   [y,count] = fread(fid,1,'uint16');

end

status = fclose(fid);

% End of READIMGDATABLOCK
```

```
function y = temptosignal(temperature, emissivity, tAmb, ...
         tAtm, humidity, distance, alpha1, ...
         alpha2, beta1, beta2, X, R, B, F, ...
         obas, L, globalGain, globalOffset)
%TEMPTOSIGNAL Calculates pixel value on the basis
%             of temperature value
```

```
%  Y = TEMPTOSIGNAL(TEMPERATURE, EMISSIVITY, TAMB,
%                   TATM, HUMIDITY, DISTANCE, ALPHA1,
%                   ALPHA2, BETA1, BETA2, X, R, B, F,
%                   OBAS, L, GLOBALGAIN, GLOBALOFFSET)
%
% TEMPERATURE: Measured temperature
% EMISSIVITY:  Emissivity of object
% TAMB:        Ambient temperature
% TATM:        Temperature of atmosphere
% HUMIDITY:    Relative humidity
% DISTANCE:    Camera-to-object distance,
%
% R, B, F, X,
% ALPHA1, ALPHA2,
% BETA1, BETA2,
% GLOBALGAIN,
% GLOBALOFFSET, L, OBAS: Calibration and adjusting
%                        parameters
%
%See also: LOADIMGHEADER, CAMERAMODEL
%
%Copyright November 2008 by Sebastian Dudzik.


tAtmCelsius = tAtm - 273.15;

h2o = humidity*exp(1.5587 + (6.939e-2).* ...
      tAtmCelsius - (2.7816e-4).*tAtmCelsius^2 + ...
      (6.8455e-7).*tAtmCelsius^3);

tau = X*exp((-1)*sqrt(distance)* ...
     (alpha1 + beta1*sqrt(h2o))) + ...
     (1-X)*exp((-1)*sqrt(distance)* ...
     (alpha2 + beta2*sqrt(h2o)));

varH = exp(B./temperature);
varH = varH - F;
objSignal = R./varH;
k1 = 1./(emissivity.*tau);
k2 = (((1 - emissivity)./emissivity).* ...
    (R./(exp(B./tAmb) - F))) + ...
    (((1 - tau)./(emissivity.*tau)).* ...
    (R./(exp(B./tAmb) - F)));
absPixel = 2*(objSignal + k2)./k1;
lFunc = (absPixel - globalOffset)./globalGain;
y = obas+(lFunc./(1 + L.*lFunc));

% End of TEMPTOSIGNAL
```

A.7 Sample MATLAB Sessions

A.7.1 Calculation of the Components of the Combined Standard Uncertainty of Object Temperature

Problem. *Calculate the components of the combined standard uncertainty of object temperature for the following measurement conditions (Table A.3) (see Section 5.2.3 and Tables 5.2 and 5.3):*

Table A.3 Measurement conditions for the sample session of calculation of combined standard uncertainty of object temperature

Input quantity	Object emissivity (ε_{ob})	Ambient temperature (T_o), K	Atmospheric temperature (T_{atm}), K	Relative humidity (ω)	Camera-to-object distance (d), m
Estimate value	0.9	293	293	0.5	1
Uncertainty range	0–30%	0–3%	0–3%	0–30%	0–30%

Value of temperature measured with camera equals 363 K. The thermal image recorded with an infrared camera was stored in file: D1017-10.img.

Solution. *Presented below is the sample MATLAB session for the solution of this problem using the m-files* `components.m` *and* `plotcomponents.m` *from Sections A.5 and A.6.*

```
>> components
****************************************
*                                      *
*  Calculation of components of the    *
*  combined standard uncertainty       *
*  of the object temperature           *
*  for FLIR ThermaCAM infrared cameras  *
*                                      *
****************************************

*** FILE NAME AND MEASURED TEMPERATURE BLOCK ***

Name of *.img recorded file
(must be in the same dir): D1017-10.img
Value of measured temperature (K):363

*** REFERENCE CONDITIONS BLOCK ***
Value of emissivity: 0.9
Value of ambient temperature (K): 293
Value of temperature of atmosphere (K): 293
Value of relative humidity : 0.5
Value of camera-to-object distance (m): 1

*** RANGES OF THE STANDARD UNCERTAINTIES OF
INPUT VARIABLES BLOCK ***
```

```
Minimum uncertainty of emissivity (%): 0
Maximum uncertainty of emissivity (%): 30
Minimum uncertainty of ambient temperature (%): 0
Maximum uncertainty of ambient temperature (%): 3
Minimum uncertainty of temperature of atmosphere (%): 0
Maximum uncertainty of temperature of atmosphere (%): 3
Minimum uncertainty of relative humidity (%): 0
Maximum uncertainty of relative humidity (%): 30
Minimum uncertainty of camera-to-object distance (%): 0
Minimum uncertainty of camera-to-object distance (%): 30
Number of simulation points: 100

*** THE DISTRIBUTIONS OF THE INPUT RANDOM VARIABLES BLOCK ***
1 - Lognormal distribution
2 - Uniform distribution

Enter type of distribution (1/2): 2
>> plotcomponents
```

The resulting plot can be compared to the component curve representing the value of measured temperature equal to 363 K in Figures 5.23a, 5.25a, 5.27a, 5.29a, 5.31a (Sections 5.3.1–5.3.5).

A.7.2 Calculation of the Combined Standard Uncertainty and the 95% Coverage Interval of the Object Temperature

Problem. *Calculate the combined standard uncertainty and the 95% coverage interval of the object temperature for following measurement conditions (Table A.4) (see Section 5.5.2 and Tables 5.7 and 5.8):*

Table A.4 Measurement conditions for the sample session of calculation of standard uncertainty and the 95% coverage interval of the object temperature

Input quantity	Object emissivity (ε_{ob})	Ambient temperature (T_o), K	Atmospheric temperature (T_{atm}), K	Relative humidity (ω)	Camera-to-object distance (d), m
Estimate value	0.9	293	293	0.5	10
Uncertainty range	0.09	9	9	0.05	1

Value of temperature measured with camera equals 363 K. The thermal image recorded with an infrared camera was stored in file: D1017-10.img.

Solution. *Presented below is the sample MATLAB session for the solution of this problem using the m-files* `coverint.m` *and* `plotresults.m` *from Sections A.5 and A.6.*

```
>> coverint
*****************************************
*                                       *
*   Calculation of coverage interval    *
*   for FLIR ThermaCAM infrared cameras *
*                                       *
*****************************************

*** FILE NAME AND MEASURED TEMPERATURE BLOCK ***

Name of *.img recorded file
(must be in the same dir): D1017-10.img

Value of measured temperature (K):363

*** REFERENCE CONDITIONS BLOCK ***

Value of emissivity: 0.9
Value of ambient temperature (K): 293
Value of temperature of atmosphere (K): 293
Value of relative humidity : 0.5
Value of camera-to-object distance (m): 10

*** STANDARD UNCERTAINTIES OF INPUT VARIABLES BLOCK ***

Standard uncertainty of emissivity: 0.09
Standard uncertainty of ambient temperature (K): 9
Standard uncertainty of temperature of atmosphere (K): 9
Standard uncertainty of relative humidity : 0.05
Standard uncertainty of camera-to-object distance (m): 1

*** RESULTS BLOCK ***

Combined standard uncertainty of object temperature
ans =
    5.6505

95% coverage interval ([tLow tHigh]): [355.063, 374.2662]
>> plotresults
```

The results can be compared to Table 5.9 and Figure 5.59 (Section 5.5.2).

A.7.3 Simulation of the Relative Combined Standard Uncertainty Versus the Correlation Coefficient of Selected Input Random Variables

Problem. *Simulate the influence of the cross-correlations between the input random variables representing object emissivity and atmospheric temperature for the following measurement conditions (Table A.5) (see Section 5.4.2 and Tables 5.4 and 5.5):*

Value of temperature measured with camera equal 363 K. The values of correlation coefficient changed from −0.99 to 0.99 in steps of 0.01. The thermal image recorded with an infrared camera was stored in file: D1017-10.img.

Table A.5 Measurement conditions for the simulation of the dependence between the relative combined standard uncertainty and correlation coefficient for selected input random variables

Input quantity	Object emissivity (ε_{ob})	Ambient temperature (T_o), K	Atmospheric temperature (T_{atm}), K	Relative humidity (ω)	Camera-to-object distance (d), m
Estimate value	0.9	293	293	0.5	50
Uncertainty value	0.09	29.3	29.3	0.05	5

Solution. *Presented below is the sample MATLAB session for the solution of this problem using the m-files* `correlations.m` *and* `plotcorrsens.m` *from Sections A.5 and A.6.*

```
>> correlations
*****************************************
*                                       *
*   Simulates the influence of the      *
*   cross-correlations between the input *
*   variables of the FLIR ThermaCAM     *
*   infrared camera model on the        *
*   combined standard uncertainty       *
*                                       *
*****************************************

*** FILE NAME AND MEASURED TEMPERATURE BLOCK ***
Name of *.img recorded file
(must be in the same dir): D1017-10.img

Value of measured temperature (K): 363

*** REFERENCE CONDITIONS BLOCK ***

Value of emissivity: 0.9
Value of ambient temperature (K): 293
Value of temperature of atmosphere (K): 293
Value of relative humidity : 0.5
Value of camera-to-object distance (m): 50

*** STANDARD UNCERTAINTIES OF INPUT VARIABLES BLOCK ***

Standard uncertainty of emissivity: 0.09
Standard uncertainty of ambient temperature (K): 29
Standard uncertainty of temperature of atmosphere (K): 29
Standard uncertainty of relative humidity : 0.05
Standard uncertainty of camera-to-object distance (m): 5

*** PARAMETERS OF CROSS-CORRELATIONS VECTOR BLOCK ***

Starting value of cross-correlation vector: -0.99
Step value into cross-correlation vector: 0.01
Ending value of cross-correlation vector: 0.99
```

```
*** THE DISTRIBUTIONS OF THE INPUT RANDOM VARIABLES BLOCK ***

1 - Lognormal distribution
2 - Uniform distribution
Enter type of distribution (1/2): 2

*** THE CROSS-CORRELATED INPUT RANDOM VARIABLES BLOCK ***
The List of the input variables' index
Index | Input variable
--------|-----------------
  1     | Emissivity
  2     | Ambient temperature
  3     | Atmosphere temperature
  4     | Relative humidity
  5     | Camera-to-object distance

Enter the index of the first
cross-correlated input variable: 1

Enter the index of the second
cross-correlated input variable: 3

>> plotcorrsens
```

The resulting plots can be compared to the curve representing the value of measured temperature equal to 363 K in Figure 5.35a (Section 5.4.2).

Appendix B

Normal Emissivities of Various Materials (IR-Book 2000, Minkina 2004)

Material	Specification	Temperature, °C	Spectrum[a]	Emittance
METALS and METAL OXIDES				
Aluminum	Foil	27	10 μm	0.04
	Foil	27	3 μm	0.09
	Vacuum deposited	20	T	0.04
	Polished	50–100	T	0.04–0.06
	Polished plate	100	T	0.05
	Anodized, black, dull	70	LW	0.95
	Anodized sheet	100	T	0.55
Aluminum bronze		20	T	0.60
Aluminum hydroxide	Powder		T	0.28
Aluminum oxide	Activated, powder		T	0.46
Aluminum oxide	Pure, powder (alumina)		T	0.16
Brass	Polished, highly	100	T	0.03
	Polished	200	T	0.03
	Sheet, rolled	20	T	0.06
Bronze	Polished	50	T	0.1
	Porous, rough	50–150	T	0.55
Chromium	Polished	500–1000	T	0.28–0.38
	Polished	50	T	0.10

(continued)

METALS and METAL OXIDES (*continued*)

Material	Specification	Temperature, °C	Spectrum[a]	Emittance
Copper	Electrolytic, polished	−34	T	0.006
	Electrolytic, carefully polished	80	T	0.018
	Pure, carefully prepared surface	22	T	0.008
	Polished, mechanical	22	T	0.015
	Polished	50–100	T	0.02
	Oxidized, black	27	T	0.78
	Oxidized to blackness		T	0.88
Copper oxide	Red, powder		T	0.70
Copper dioxide	Powder		T	0.84
Gold	Polished	130	T	0.018
	Polished, carefully	200–600	T	0.02–0.03
	Polished, highly	100	T	0.02
Iron, cast	Polished	38	T	0.21
	Polished	40	T	0.21
	Unworked	900–1100	T	0.87–0.95
	Ingots	1000	T	0.95
Iron and steel	Electrolytic	22	T	0.05
	Electrolytic	100	T	0.05
	Electrolytic, carefully polished	175–225	T	0.05–0.06
	Electrolytic	260	T	0.07
	Polished	400–1000	T	0.14–0.38
	Oxidized	200–600	T	0.80
Iron tinned	Sheet	24	T	0.064
Iron galvanized	Sheet	92	T	0.07
	Sheet, burnished	30	T	0.23
	Sheet, oxidized	20	T	0.28
	Heavily oxidized	70	SW	0.64
	Heavily oxidized	70	LW	0.85
Lead	Unoxidized, polished	100	T	0.05
	Shiny	250	T	0.08
	Oxidized at 200 °C	200	T	0.63
Lead red		100	T	0.93
Magnesium		22	T	0.07
	Polished	20	T	0.07
Magnesium powder			T	0.86

METALS and METAL OXIDES *(continued)*

Material	Specification	Temperature, °C	Spectrum[a]	Emittance
Molybdenum		600–1000	T	0.08–0.13
	Filament	700–2500	T	0.1–0.3
		1500–2200	T	0.19–0.26
Nichrome	Rolled	700	T	0.25
	Sandblasted	700	T	0.70
	Wire, clean	500–1000	T	0.71–0.79
	Wire, clean	50	T	0.65
	Wire, oxidized	50–500	T	0.95–0.98
Nickel	Electrolytic	22	T	0.04
	Electrolytic	38	T	0.06
Nickel oxide		500–650	T	0.52–0.59
		1000–1250	T	0.75–0.86
Platinum		17	T	0.016
		22	T	0.03
		100	T	0.05
		260	T	0.06
		538	T	0.10
	Ribbon	900–1100	T	0.12–0.17
Silver	Polished	100	T	0.03
	Pure, polished	200–600	T	0.02–0.03
Stainless steel	Type 18-8, buffed	20	T	0.16
	Type 18-8, oxidized at 800 °C	60	T	0.85
	Sheet, polished	70	SW	0.18
	Sheet, polished	70	LW	0.14
	Sheet, untreated, somewhat scratched	70	SW	0.30
	Sheet, untreated, somewhat scratched	70	LW	0.28
	Rolled	700	T	0.45
	Sandblasted	700	T	0.70
Tin	Burnished	20–50	T	0.04–0.06
	Tin-plated sheet iron	100	T	0.07
Titanium	Polished	200	T	0.15
	Polished	500	T	0.20
	Oxidized at 540 °C	200	T	0.40
Tungsten		200	T	0.05
		600–1000	T	0.1–0.16
		1500–2200	T	0.24–0.31
	Filament	3300	T	0.39

(continued)

METALS and METAL OXIDES *(continued)*

Material	Specification	Temperature, °C	Spectrum[a]	Emittance
Zinc	Polished	200–300	T	0.04–0.05
	Oxidized at 400 °C	400	T	0.11
	Sheet	50	T	0.20
	Oxidized surface	1000–1200	T	0.50–0.60

OTHER MATERIALS

Material	Specification	Temperature, °C	Spectrum[a]	Emittance
Asbestos	Powder		T	0.40–0.60
	Fabric		T	0.78
	Floor tile	35	SW	0.94
	Board	20	T	0.96
	Slate	20	T	0.96
	Paper	40–400	T	0.93–0.95
Asphalt paving		4	LLW	0.967
Brick	Sillimanite, 33% SiO_2, 64% Al_2O_3	1500	T	0.29
	Refractory, magnesite	1000–1300	T	0.38
	Refractory, corundum	1000	T	0.46
	Refractory, weakly radiating	500–1000	T	0.65–0.75
	Waterproof	17	SW	0.87
	Red, rough	20	T	0.88–0.93
	Red, common	20	T	0.93
	Masonry	35	SW	0.94
	Masonry, plastered	20	T	0.94
Carbon	Candle soot	20	T	0.95
	Lampblack	20–400	T	0.95–0.97
Chipboard	Untreated	20	SW	0.90
Clay	Fired	70	T	0.91
Cloth	Black	20	T	0.98
Concrete		20	T	0.92
	Dry	36	SW	0.95
	Rough	17	SW	0.97
	Walkway	5	LLW	0.974
Ebonite			T	0.89
Emery	Coarse	80	T	0.85
Enamel		20	T	0.9
	Lacquer	20	T	0.85–0.95
Fiberboard	Hard, untreated	20	SW	0.85
	Porous, untreated	20	SW	0.85
	Masonite	70	LW	0.88

OTHER MATERIALS (*continued*)

Material	Specification	Temperature, °C	Spectrum[a]	Emittance
Granite	Polished	20	LLW	0.849
	Rough	21	LLW	0.879
	Rough, four different samples	70	SW	0.95–0.97
Gypsum		20	T	0.8–0.9
Ice: see Water				
Lacquer	Aluminum on rough surface	20	T	0.4
	Bakelite	80	T	0.83
	Black, matte	100	T	0.97
	Black, dull	40–100	T	0.96–0.98
	White	40–100	T	0.8–0.95
Leather	Tanned		T	0.75–0.80
Lime			T	0.3–0.4
Mortar		17	SW	0.87
	Dry	36	SW	0.94
Oil, lubricating	Film on Ni base: Ni base only	20	T	0.05
	: 0.025 mm film	20	T	0.27
	: 0.050 mm film	20	T	0.46
	: 0.125 mm film	20	T	0.72
	: thick coating	20	T	0.82
Paint	Aluminum, various ages	50–100	T	0.27–0.67
	Cadmium yellow		T	0.28–0.33
	Chrome green		T	0.65–0.70
	Cobalt blue		T	0.7–0.8
	Oil	17	SW	0.87
	Oil, grey gloss	20	SW	0.96
	Plastic, black	20	SW	0.95
Paper	Yellow		T	0.72
	Red		T	0.76
	Blue, dark		T	0.84
	Green		T	0.85
	Black		T	0.90
	Coated with black lacquer		T	0.93
	White	20	T	0.7–0.9
	White bond	20	T	0.93
	Black, dull	70	SW	0.86
	Black, dull	70	LW	0.89
	Black, dull		T	0.94

(*continued*)

OTHER MATERIALS (*continued*)

Material	Specification	Temperature, °C	Spectrum[a]	Emittance
Plaster		17	SW	0.86
	Rough coat	20	T	0.91
	Plasterboard, untreated	20	SW	0.90
Plastic	PVC, plastic floor, dull, structured	70	SW	0.94
	PVC, plastic floor, dull, structured	70	LW	0.93
Porcelain	White, shiny		T	0.70–0.75
	Glazed	20	T	0.92
Rubber	Hard	20	T	0.95
	Soft, gray, rough	20	T	0.95
Sand			T	0.60
		20	T	0.90
Sandstone	Polished	19	LLW	0.909
	Rough	19	LLW	0.935
Skin	Human	32	T	0.98
Slag	Boiler	0–100	T	0.97–0.93
	Boiler	200–500	T	0.89–0.78
Soil	Dry	20	T	0.92
	Saturated with water	20	T	0.95
Stucco	Rough, lime	10–90	T	0.91
Styrofoam	Insulation	37	SW	0.60
Tar			T	0.79–0.84
	Paper	20	T	0.91–0.93
Tile	Glazed	17	SW	0.94
Wallpaper	Slight pattern, light grey	20	SW	0.85
	Slight pattern, red	20	SW	0.90
Varnish	Flat	20	SW	0.93
	On oak parquet floor	70	SW	0.90
	On oak parquet floor	70	LW	0.90–0.93
Water	Snow		T	0.8
	Snow	−10	T	0.85
	Layer > 0.1 mm thick	0–100	T	0.95–0.98
	Distilled	20	T	0.96
	Ice, smooth	−10	T	0.96
	Ice, smooth	0	T	0.97
	Ice, covered with heavy frost	0	T	0.98
	Frost crystals	−10	T	0.98

OTHER MATERIALS *(continued)*

Material	Specification	Temperature, °C	Spectrum[a]	Emittance
Wood	Ground		T	0.5–0.7
	Planed	20	T	0.8–0.9
	Planed oak	20	T	0.90
	White, damp	20	T	0.7–0.8
	Planed oak	70	SW	0.77
	Planed oak	70	LW	0.88

[a]T, total spectrum $(0-\infty)$ μm; SW, 2–5 μm; LW, 8–14 μm; LLW, 6.5–20 μm.

Bibliography

References

Astarita T., Cardone G., Carlomagno G.M. and Meola C. (2000) 'A survey on infrared thermography for convective heat transfer measurements', *Optics & Laser Technology*, Vol. 32, pp. 593–610.

Bąbka R. and Minkina W. (2001) 'Simulation research of usability of the mean square metric to analysis of temperature fields recorded with infrared system', Proceedings of XI Symposium 'Modelowanie i Symulacja Systemów Pomiarowych (Modeling and Simulation of Measurement Systems)' (MiSSP'2001), 17–21 September, Krynica Górska, pp. 129–136 (in Polish).

Bąbka R. and Minkina W. (2002a) 'Influence of calibration of an infrared camera on accuracy of sub-pixel edge detection of thermal objects', *Measurement, Automation and Monitoring*, Vol. 48, No. 4, pp. 11–13 (in Polish).

Bąbka R. and Minkina W. (2002b) 'Sub-pixel edge location in thermal images using a meansquare measure', Proceedings of 7th International Conference on Infrared Sensors & Systems (IRS^2 2002), ed. G. Gerlach, 14–16 May, Erfurt, pp. 213–218.

Bąbka R. and Minkina W. (2002c) 'Application of the statistical hypothesis method to identification of thermal anomalies in electric devices', Proceedings of V National Conference 'Termografia i Termometria w Podczerwieni (Thermography and Thermometry in Infrared)' (TTP'2002), ed. B. Wiecek,14–16 November, Ustroń, pp. 353–358 (in Polish).

Bąbka R. and Minkina W. (2003a) 'Analysis of a model for determination of edge locations with sub-pixel accuracy using an infrared thermovision system', Proceedings of XIII Symposium 'Modelowanie i Symulacja Systemów Pomiarowych (Modeling and Simulation of Measurement Systems)' (MiSSP'2003), 8–11 September, Kraków, pp. 197–204 (in Polish).

Bąbka R. and Minkina W. (2003b) 'Estimation of development of thermal defects in electric devices using a thermovision system', Proceedings of Central European V Conference 'Metody i Systemy Komputerowe w Automatyce i Elektrotechnice (Computer-aided Methods and Systems in Automatics and Electronics)' (MSKAE'2003), 16–17 September, Częstochowa – Poraj, pp. 40–43 (in Polish).

Bayazitoğlu Y. and Özişik M.N. (1988) *Elements of heat transfer*, McGraw-Hill, New York.

Bernhard F.(ed.) (2003) *Technische Temperaturmessung – physikalische und meβtechnische Grundlagen, Sensoren und Meβverfahren, Meβfehler und Kalibrierung*, Springer-Verlag, Berlin.

Bielecki Z. and Rogalski A. (2001) *Detection of optical signals*, WNT, Warsaw (in Polish).

Breiter R., Cabanski W., Koch R., Mauk K.-H., Rode W., Ziegler J., Eberhardt K., Oelmaier R., Walther M. and Schneider H. (2000) 'Large format focal plane array IR detection modules and digital signal processing technologies at AIM', Proceedings of 6th Conference on Infrared Sensors & Systems (IRS^2 2000), ed. G. Gerlach, 9–11 May, Erfurt, pp. 25–30.

Breiter R., Cabanski W., Koch R., Mauk K.-H., Rode W., Ziegler J., Walther M., Schneider H. and Oelmaier R. (2002) '3rd generation focal plane array IR detection modules at AIM', Proceedings of 7th International Conference on Infrared Sensors & Systems (IRS^2 2002), ed. G. Gerlach,14–16 May, Erfurt, pp. 19–26.

Chrzanowski K. (2000) *Non-contact thermometry – measurement errors*, SPIE, Vol. 7, Polish Chapter, Warsaw.

CIPM (1981) 'BIPM Proceedings - Verb. Com. Int. Poids et Mesures 49' (in French).

CIPM (1986) 'BIPM Proceedings - Verb. Com. Int. Poids et Mesures 54' (in French).

Cox M.G., Dainton M.P. and Harris P.M. (2001) 'Best Practice Guide No. 6. Uncertainty and statistical modelling', Technical Report, National Physical Laboratory, Teddington, UK.

Danjoux R. (2001) 'The evolution in spatial resolution', *InfraMation Magazine – itc-i*, Vol. 2, No. 12, pp. 1–3.

De Mey G. (1989) 'Thermographic measurements of hybrid circuits', Proceedings of 7th European Hybrid Microelectronics Conference, Hamburg, pp. 1–7.

De Mey G. and Wiecek B. (1998) 'Application of thermography for microelectronic design and modelling', Proceedings of Quantitative InfraRed Thermography (QIRT'98), ed. D. Balageas *et al.*, 7–10 September, Łódź, pp. 23–27.

DeWitt D.P. (1983) 'Inferring temperature from optical radiation measurements', *Proceedings of the SPIE*, Vol. 446, pp. 226–233.

Dudzik S. (2000) 'A measurement system for digital processing of thermal images', Master's thesis, Częstochowa University of Technology (in Polish).

Dudzik S. (2003) 'TermoLab – a digital system for thermal images processing with universal matrix interface', Proceedings of XXXV Intercollegiate Conference of Metrologists (MKM'03), 8–11 September, Kraków, pp. 95–98 (in Polish).

Dudzik S. (2005) 'Analysis of influence of correlation coefficient between measurement model input variables on uncertainty of temperature determination with an infrared camera', Proceedings of XXXVII Intercollegiate Conference of Metrologists (MKM'05), 5–7 September, Zielona Góra, pp. 195–203 (in Polish).

Dudzik S. (2007) 'Evaluation of central heating radiators power using computer-aided thermography and tuned heat transfer models', Doctoral thesis, Institute of Electronics and Control Systems, Częstochowa (in Polish).

Dudzik S. (2008) 'Calculation of the heat power consumption in the heat exchanger using artificial neural network', Proceedings of 9th Conference on Quantitative Infrared Thermography (QIRT'2008), 2–5 July, Kraków, pp. 55–60.

Dudzik S. and Minkina W. (2002) 'Thermovision measurements as a source of data for calculation of radiators heat power using a convective-radiation heat transfer model', Proceedings of V National Conference 'Termografia i Termometria w Podczerwieni (Thermography and Thermometry in Infrared)' (TTP'2002), ed. B. Wiecek,14–16 November, Ustroń-Jaszowiec, pp. 181–186 (in Polish).

Dudzik S. and Minkina W. (2007) 'Assessment of influence of input variables correlation of a thermovision measurement model on temperature evaluation accuracy for low emissivity objects', *Measurement, Automation and Monitoring*, Vol. 57, No. 2, pp. 29–32 (in Polish).

Dudzik S. and Minkina W. (2008a) 'The application of artificial neural network for calculation of heat power consumption, using infrared system', Proceedings of 10th International Conference of Infrared Sensors & Systems (IRS^2 2008)', ed. G. Gerlach, 6–7 May, Nürnberg, pp. 317–321.

Dudzik S. and Minkina W. (2008b) 'Application of the numerical method for the propagation of distributions to the calculation of coverage intervals in the thermovision measurements', Proceedings of 9th Conference on Quantitative Infrared Thermography (QIRT'2008), 2–5 July, Kraków, pp. 179–184.

Fuller, W.A. (1987) *Measurement error models*, John Wiley & Sons, Inc., New York.

Gajda J. and Szyper M. (1998) *Modeling and simulation investigation of measurement systems*, Jartek Publisher, Kraków (in Polish).

Gaussorgues, G. (1994) *Infrared thermography* (Microwave Technology Series 5), Chapman & Hall, London.

Gerashenko O.A., Gordow A.N., Eremina A.K., Lakh W.I., Lucyk Y.T., Pucylo W.I., Stadnyk B.I. and Yaryshew N.A. (1989) *Tiempieratiurne izmierienia – sprawoenyk*, Naukova Dumka, Kiev.

Glückert U. (1992) *Erfassung und Messung von Wärmestrahlung*, Franzis-Verlag, München.

Guide (1995) 'Guide to the Expression of Uncertainty in Measurement', International Organization for Standarization (ISO), Geneva (CH) - Corrected and reprinted/TAG.

Guide (2004) 'Guide to the Expression of Uncertainty in Measurement. Supplement 1. Numerical Methods for the Propagation of Distributions', Document of International Bureau of Weights and Measures.

Hamrelius T. (1991) 'Accurate temperature measurement in thermography. An overview of relevant features, parameters and definitions', *Proceedings of the SPIE*, Vol. 1467, pp. 448–457.

Hartmann J., Gutschwager B., Fischer J. and Hollandt J. (2002) 'Calibration of thermal cameras for temperature measurements using black body radiation in the temperature range –60°C up to 3000°C', Proceedings of 7th International Conference on Infrared Sensors & Systems (IRS^2 2002), ed. G. Gerlach,14–16 May, Erfurt, pp. 119–124.

Herschel J.F.W. (1830) *Preliminary discourse on the study of natural philosophy*, Longman Rees. Grme, Brown. & Green, London (Polish edition: *Wstęp do badań przyrodniczych*, PWN, Warszawa, 1955).

Herschel W. (1800a) 'Investigation of the powers of the prismatic colours to heat and illuminate objects; with remarks, that prove the different refrangibility of radiant heat. To which is added, an inquiry into the method of viewing the sun advantageously, with telescopes of large apertures and high magnifying powers', *Philosophical Transactions of the Royal Society of London*, Vol. 90, part II, pp. 255–283.

Herschel W. (1800b) 'Experiments on the refrangibility of the invisible rays of the sun', *Philosophical Transactions of the Royal Society of London*, Vol. 90, part II, pp. 284–292.

Herschel W. (1800c) 'Experiments on the solar, and on the terrestrial rays that occasion heat; with a comparative view of the laws to which light and heat, or rather the rays which occasion them, are subject, in order to determine whether they are the same, or different – part I', *Philosophical Transactions of the Royal Society of London*, Vol. 90, part II, pp. 293–326.

Herschel W. (1800d) 'Experiments on the solar, and on the terrestrial rays that occasion heat; with a comparative view of the laws to which light and heat, or rather the rays which occasion them, are subject, in order to determine whether they are the same, or different – part II', *Philosophical Transactions of the Royal Society of London*, Vol. 90, part III, pp. 437–538.

Hudson R.D. (1969) *Infrared systems engineering*, John Wiley & Sons, Inc., New York.

Hutton, D.V. (2003) *Fundamentals of finite element analysis*, McGraw-Hill, New York.

IR-Book (2000) *FLIR Training Proceedings, Level II*, Infrared Training Center – International.

Janna W.S. (2000) *Engineering heat transfer*, CRS Press, Washington, DC.

Kaplan H. (2000) 'Infrared spectral bands – the importance of color in the infrared', Proceedings of InfraMation Conference, Vol. 1, ed. G. Orlove, 24–27 September, Orlando, FL.

Kreith F. (2000) *The CRC handbook of thermal engineering*, CRC Press, Boca Raton, FL.

Lieneweg F. (ed.) (1976) *Handbuch der technischen Temperaturmessung*, Friedr. Vieweg & Sohn, Braunschweig.

Linhart J. and Linhart S. (2002) 'Temperature measurements in the glass melt', *Glass Technology*, Vol. 43, No. 5 (October), pp. 1–5.

Machin G. and Chu B. (2000) 'High quality blackbody sources for infrared thermometry and thermography between −40°C and 1000°C', *Imaging Science*, Vol. 48, pp. 15–22.

Machin G., Simpson R. and Broussely M. (2008) 'Calibration and validation of thermal imagers', Proceedings of Quantitative InfraRed Thermography (QIRT'2008), ed. B. Wiecek, 2–5 July, Krakow, pp. 555–562.

Madding R.P. (1999) 'Emissivity measurement and temperature correction accuracy considerations', *Proceedings of the SPIE*, Vol. 3700, pp. 393–401.

Madding R.P. (2000) 'Common misconceptions in infrared thermography condition based maintenance applications', Proceedings of InfraMation Conference, Vol. 1, ed. G. Orlove, 24–27 September, Orlando, FL.

Madura H. *et al.* (2001) 'Two-color pyrometer for non-contact temperature measurements with increased data acquisition rate', Report from research project KBN 8T10C 045 99C/4611 (in Polish).

Madura H. *et al.* (2004) *Thermovision measurements in practice*, PAK Publishers Agenda, Warsaw (in Polish).

Maldague X.P. (2001) *Theory and practice of infrared technology for nondestructive testing*, Wiley Interscience, New York.

Marshall S.J. (1981) 'We need to know more about infrared emissivity', *Proceedings of the SPIE*, Vol. 0313, pp. 119–127.

MATLAB (2005a) *MATLAB R13 User's Guide*, The MathWorks Inc., Natick, MA.

MATLAB (2005b) 'Simulation dependent random variables using copulas', *MATLAB 7.0.4*, The MathWorks Inc., Natick, MA.

McGee T.D. (1998) *Principles and methods of temperature measurement*, John Wiley & Sons, Inc., New York.

Michalski L., Eckersdorf K. and McGhee J. (1991) *Temperature measurement*, John Wiley & Sons, Ltd, Chichester.

Michalski L., Eckersdorf K. and Kucharski J. (1998) *Thermometry – methods and instruments*, Łódź University of Technology Publishers, Łódź (in Polish).

Minkina W. (1992) 'Non-linear models of temperature sensor dynamics', *Sensors & Actuators A: Physical*, Vol. 30, No. 3, pp. 209–214.

Minkina W. (1994) 'Space discretization in solving chosen problems of unsteady heat conductivity by means of electric modeling', *Experimental Technique of Physics (ETP)*, Vol. 40, No. 2, pp. 283–302.

Minkina W. (1995) 'On some singularities in space discretization while solving the problems of unsteady heat conduction', *Experimental Technique of Physics (ETP)*, Vol. 41, No. 1, pp. 37–54.

Minkina W. (1999) 'Theoretical and experimental identification of the temperature sensor unit step response non-linearity during air temperature measurement', *Sensors & Actuators A: Physical*, Vol. 78, No. 2–3, pp. 81–87.

Minkina W. (2001) 'Fundamentals of thermovision measurements, part III – metrological problems, interpretation of results', *Measurement, Automation and Monitoring*, Vol. 47, No. 11, pp. 5–8 (in Polish).

Minkina W. (2003) 'Thermometry – fundamental problems of thermovision measurements', Proceedings of XXXV Międzyuczelnianej Konferencji Metrologów (Intercollegiate Conference of Metrologists)' (MKM'2003), 8–11 September, Kraków, pp. 77–86 (in Polish).

Minkina W. (2004) *Thermovision measurements – methods and instruments*, Publishing Office of Częstochowa University of Technology, Częstochowa (in Polish).

Minkina W. and Bąbka R. (2002) 'Influence of components of the error of method on error of temperature indication on the basis of the ThermaCAM PM595 infrared camera measurement model', Proceedings of V National Conference 'Termografia i Termometria w Podczerwieni (Thermography and Thermometry in Infrared)' (TTP'2002), ed. B. Wiecek,14–16 November, Ustroń-Jaszowiec, pp. 339–344 (in Polish).

Minkina W. and Chudzik S. (2003) 'Determination of thermal parameters of heat-insulating materials using artificial neural networks', *Quarterly: Metrology and Measurement Systems*, Vol. X, No. 1, pp. 33–49.

Minkina W. and Dudzik S. (2005) 'Simulation analysis of uncertainty of processing algorithm of the ThermaCAM PM 595 infrared camera measurement path', Proceedings of III Symposium 'Metrologiczne Właściwości Programowanych Przetworników Pomiarowych (Metrological Properties of Programmable Measuring Transducers)' (MWPPP'2004), 22–23 November, Gliwice, pp. 173–185 (in Polish).

Minkina W. and Dudzik S. (2006a) 'Simulation analysis of uncertainty of infrared camera measurement and processing path', *Measurement*, Vol. 39, No. 8, pp. 758–763.

Minkina W. and Dudzik S. (2006b) 'Simulation analysis of sensitivity of thermovision temperature measurement model', *Measurement, Automation and Monitoring*, Vol. 52, No. 9, pp. 56 59 (in Polish).

Minkina W. and Gryś S. (2002) 'Application of adaptive signal processing in error compensation of transient temperature measurements', *Quarterly: Metrology and Measurement Systems*, Vol. IX, No. 2, pp. 123–137.

Minkina W. and Gryś S. (2005) 'Dynamics of contact thermometric sensors with electric output and methods of its improvement', *Quarterly: Metrology and Measurement Systems*, Vol. XII, No. 4, pp. 371–391.

Minkina W. and Wild. W. (2004) 'Selected metrological problems in thermovision measurements', International Conference on Technical and Economic Aspects of Modern Technology Transfer in Context of Integration with European Union (TEAMT'2004), 25–27 November, Złoty Potok, near Częstochowa, pp. 363–371.

Minkina W., Rutkowski P. and Wild W. (2000) 'Fundamentals of thermovision measurements part I – the essence and history of thermovision, part II – contemporary designs of thermovision systems, errors of method', *Measurement, Automation and Monitoring*, Vol. 46, No. 1, pp. 7–14 (in Polish).

Minkina W., Bąbka R. and Wild W. (2001) 'Assessment of application of the meansquare measure for thermal object edge detection in the thermovision images', *Measurement, Automation and Monitoring*, Vol. 47, No. 11, pp. 9–12 (in Polish).

Minkina W., Dudzik S. and Wild W. (2002) 'TermoLab – a digital system for thermal images, processing with universal matrix interface', *Measurement, Automation and Monitoring*, Vol. 48, No. 4, pp. 9–10 (in Polish).

Minkina W., Dudzik S., Bąbka R. and Wild W. (2003) 'TermoLab – a system of analysis and visualization of thermal images recorded with ThermaCAM PM 575/595 infrared cameras', *Informatyka Teoretyczna i Stosowana (Theoretical and Applied Informatics)*, Vol. 3, No. 4, pp. 99–114 (in Polish).

Nelles O. (2001) *Nonlinear system identification. From classical approaches to neural networks and fuzzy models*, Springer-Verlag, Heidelberg.

Nowakowski A. (2001) *Advances in thermography – medical applications*, Gdańskie Publisher, Gdańsk (in Polish).

Orlove G.L. (1982) 'Practical thermal measurement techniques', *Proceedings of the SPIE*, Vol. 371, pp. 72–81.

Orzechowski T. (2002) 'Techniques of thermovision measurements in machine diagnostic', *Measurement, Automation and Monitoring*, Vol. 48, No. 4, pp. 18–20 (in Polish).

Özisik M.N. (1973) *Radiative transfer and interactions with conduction and convection*, John Wiley & Sons, Inc., New York.

Özisik N. (1994) *Finite difference methods in heat transfer*, CRC Press, Boca Raton, FL.

Pręgowski P. (2001) 'Spectral analysis of radiant signals in processes of tele-thermodetection', *Proceedings of the SPIE*, Vol. 4360, pp. 1–12.

Pręgowski P. and Świderski W. (1996) 'Experimental determination of the atmosphere – based on thermographic measurements', Proceedings of 50 Eurotherm Seminar, Quantitative Infrared Thermography (QIRT 1996), ed. D. Balageas *et al.*, Stuttgart (RFN), pp. 363–367.

Quinn T.J. (1983) *Temperature*, Academic Press, London.

QWIP Seminar (2000) Proceedings of Seminar on QWIP Detector, 7 June, Santahamina Island, Finland.

Ricolfi T. and Barber R. (1990) 'Radiation thermometers', in *Sensors – a comprehensive survey* (4 vols), ed. W. Göpel *et al.*, Vol. 4, *Thermal sensors*, ed. T. Ricolfi and J. Scholz, VCH, Weinheim, New York.

Rogalski A. (2000) *Infrared detectors*, Gordon and Breach, Amsterdam.

Rogalski A. (2003) 'Quantum well photoconductors in infrared detector technology', *Journal of Applied Physics*, Vol. 93, No. 8, pp. 4355–4391.

Rudowski G. (1978) *Thermovision and its application*, Wkit Publishers, Warsaw (in Polish).

Sala A. (1993) *Radiative heat transfer*, PWN, Warsaw (in Polish).

Satterthwaite F.E. (1941) 'Synthesis of variance', *Psychometrika*, No. 6, pp. 309–316.

Saunders P. (1999) 'Reflection errors in industrial radiation thermometry', Proceedings of 7th International Symposium on Temperature and Thermal Measurements in Industry and Science (TEMPMEKO'1999 TC 12), ed. J.F. Dubbeldam and M.J. de Groot,1–3 June, Delft, pp. 631–636.

Schael U. and Rothe H. (2002) 'Einsatz eines augensicheren, laserbasierten Meßsystems mit 1574nm Wellenlänge zur Objektdetektion im km-Bereich', *Technisches Messen*, Vol. 69, No. 9, pp. 381–389.

Schuster N. and Kolobrodov V. (2000) *Infrarotthermographie*, Wiley-VCH, Berlin.

Siegel R. and Howell J.R. (1992) *Thermal radiation heat transfer*, 3rd edition, Taylor & Francis, New York.

Skubis T. (2003) *Elaboration of measurements' results – examples*, Politechnika Śląska Publisher, Gliwice (in Polish).

Söderström T. and Stoica P. (1994) *System identification*, Prentice Hall International, Harlow.

Stahl K. and Miosga G. (1986) *Infrarottechnik*, Hüthig-Verlag, Heidelberg.

Taylor J.R. (1997) *An introduction to error analysis. The study of uncertainties in physical measurements*, 2nd edition, University Science Books, Sausalito, CA.

Tissot J.L., Bertrand F., Vilain M. and Yon J.J. (1999) 'Uncooled infrared focal plane arrays – technical trends and LETI/LIR microbolometer development', Proceedings of 5th AITA International Workshop on Advanced Infrared Technology and Applications, ed. E. Grinzato *et al.*, 29–30 September, Venice, pp. 57–63.

TOOLKIT IC2 Dig. 16 'Developers Guide 1.00', AGEMA 550/570. (Reserved materials of FLIR Company, made available to the authors.)

Touloukian Y.S. and DeWitt D.P. (1970) *Thermophysical properties of matter: thermal radiative properties – metallic elements and alloys*, Vol. 7, IFI/Plenum Data Corporation, New York and Washington, DC.

Touloukian Y.S. and DeWitt D.P. (1972) *Thermophysical properties of matter: thermal radiative properties – nonmetallic solids*, Vol. 8, IFI/Plenum Data Corporation, New York and Washington, DC.

Touloukian Y.S., DeWitt D.P. and Hernicz R.S. (1972) *Thermophysical properties of matter: Thermal radiative properties – coatings*, Vol. 9, IFI/Plenum Data Corporation, New York and Washington, DC.

VIM (1993) *International vocabulary of basic and general terms in metrology*, 2nd edition, International Organization for Standardization, Geneva.

Wallin B. (1994) 'Temperature measurement on and inside lamps', *Proceedings of the SPIE*, Vol. 2245, pp. 241–251.

Walther L. and Gerber D. (1983) *Infrarotmeßtechnik*, Verlag Technik, Berlin.

Wark K. Jr (1988) *Thermodynamics*, McGraw-Hill, New York.

Welch, B.L. (1936) 'The specification of rules for rejecting too variable a product, with particular reference to an electric lamp problem', *Journal of the Royal Statistical Society*, Suppl. 3, pp. 29–48.

Wiecek B. (1999) 'Heat transfer modelling and IR measurements of electronic devices in enclosures', Proceedings of 5th AITA International Workshop on Advanced Infrared Technology and Applications, ed. E. Grinzato *et al.*, 29–30 September, Venice, pp. 272–278.

Wiecek B., De Baetselier E. and De Mey G. (1998) 'Active thermography application for solder thickness measurement in surface mounted device technology', *Microelectronics Journal*, Vol. 29, pp. 223–228.

Wolfe W.L. and Zissis G.J. (1978) *The infrared handbook*, Office of Naval Research, Washington, DC.

Standards Cited in the Text

ASTM E 1213 Minimum resolvable temperature difference (MRTD)

ASTM E 1311 Minimum detectable temperature difference (MDTD)

ASTM E 1316 Section J. Terms
ISO 31/VI Quantities and units of light and related electromagnetic radiations

Web Addresses Cited in the Text

FLIR SYSTEMS AB: www.flir.com, www.flir.com.pl, www.infraredtraining.com
Główny Urząd Miar – Poland: www.gum.gov.pl
Raytheon Infrared: www.raytheon.com
VIGO System SA: www.vigo.com.pl

Manufacturer's Web Addresses

InfraTec: http://www.infratec.de
IRCAM: http://www.ircam.de, http://www.ircam.de/startseite/startseite_d.php
IRTech: http://www.irtech.com.pl
LAND Infrared: www.landinst.com
Raytek Corporation: www.raytek.com
ThermoSensorik: www.thermosensorik.de

Index

Infrared Thermography: Errors and Uncertainties Waldemar Minkina and Sebastian Dudzik